JOURNEY TO CHERNOBYL

ALSO BY GLENN ALAN CHENEY

Nuclear Proliferation: Problems and Promises (1996)

Nuclear Testing: The Atomic War on American Soil (1996)

El Salvador, Country in Crisis (1990)

The Mariana Scouts in Amazonia (1987)

The Mariana Scouts in the Valley of Spirits (1986)

Responsibility (1985)

Revolution in Central America (1985)

Mineral Resources (1984)

The Amazon (1984)

Mohandas Gandhi (1983)

Television in American Society (1983)

JOURNEY TO CHERNOBYL:

ENCOUNTERS IN A RADIOACTIVE ZONE

Glenn Alan Cheney

Published by Academy Chicago Publishers
An imprint of Chicago Review Press Incorporated
814 North Franklin Street
Chicago, Illinois 60610
ISBN 978-0-89733-552-2

The Library of Congress has cataloged the hardcover edition as
follows:
Cheney, Glenn Alan.
Journey to Chernobyl: encounters in a radioactive zone/
Glenn Alan Cheney.
p. cm.
ISBN 0-89733-418-3
1. Chernobyl Nuclear Accident, Chornobyl', Ukraine,
1986–Social aspects. 2. Radiation victims–Ukraine. I. Title.
HV623 1986.U38C44 1995 95-32629
363.17'99'0947714—dc20 CIP

Printed in the United States of America

TABLE OF CONTENTS

CHAPTER ONE

Boots

December 3, 1991

The Aeroflot crew argues at the hatch with mannerisms and emphases possible only in a foreign language. They have a problem with paperwork. The plane sits dockside for an hour. Half a dozen gringos and I, all "UN Experts" bound for St. Petersburg to shoot the breeze on Free Economic Zones, sit elbow-to-elbow in seven seats until we figure out that the rest of Beezneez Class is all but empty. We're it.

The stewardesses are just what you'd expect in a joke about a Russian airline: hulking, crudely made up — babushkas of the future. The plane takes off with a ferocious roar, shuddering with effort as it rises into the night and tilts to the east. The cabin fills with the scream of heavy wind and a persistent whine like high-speed gears. I'm scared.

Three hours later, we land. Ireland already? The

fellow next to me, an Indian chap named Rao, says, "It's not as far as you'd think." Maybe so. But it sure doesn't look like Ireland out there. Sidewinders of snow writhe across the ice-patched runway. It's too dark to read the letters on the side of the terminal. Rao gives me a look that either confirms or doubts that it snows in Ireland.

As we clomp down the aluminum stairs into the baggage compartment, the stewardess hands us each a chit for a "Free Beverage." Then we clomp down to the icy runway and the frostbit night. It's a crisp, quick hustle to the terminal building. The snow crunches under our shoes. Everybody but me is in street shoes. I'm glad my hiking boots didn't fit in my suitcases. We all wish we'd worn overcoats. Still walking, we look back to see what an IL-86 looks like. Despite the dramatic runway spotlights and the opaque black of the sky behind, it's your basic jet: huge, squat, stubby, *serious,* with four basic cast-iron engines on the wings, no frills attached. It was obviously not built with miles per gallon in mind. That's why we're still in North America and walking across a runway in the middle of the night.

It turns out we're in Gander, Newfoundland. The free beverage is a can of Pepsi. Everybody trades in his chits, then heads for the Duty Free shop or hangs around the lobby joking about Russia. The terminal is a cross between a Quonset hut and a woodsy lodge. On one vast wall, clocks show the hours in all the Canadian time zones. Everywhere from Vancouver to Quebec it's half-past the hour. In Gander it's 10:15.

Back on the plane, we knock down some vodka with a midnight breakfast. It beats the mineral water, which tastes of sewage and seawater. There is also a cloudy brown wine which tastes sweet and homemade. Lunch: pale, mushy peas from a can, slices of old, old beets, crunchy noodles, obvious leftovers. For dessert, lacerated apples dented with bruises, served by a woman who looks as if she lives on whatever we don't eat. She isn't a regular stewardess. I think she just comes up out of the baggage compartment to serve us apples from a tray. It's her job.

Rao warns me about the bathroom. I haven't been in there yet because I don't feel like putting my boots on. I figure I can hold it till Ireland.

* *

The Pribaltiskaya Hotel is a special economic zone in which only hard currency is good. With it you can buy anything from Russian women to Miller Lite. The hookers, strictly nocturnal, are well behaved, even shy. They sit in the bars in tight jeans or short skirts, avoiding eye contact. As employees of the Russian mafia, they're let in on a bribe and allowed to stay on good behavior. Somebody told me they charge — or maybe it was "probably charge" — a hundred dollars for the night. A lot of that is baksheesh that gets distributed to everyone from the doorman, I assume, to the hotel manager. Just as in the land of the free and the home of the brave, it's the middle-pimp who makes the most. Somebody else told me that out there in the real world — if you can call Russia real — you can get a "nice one" for five smackeroos, about one-third what a university professor makes in a month.

The food in the hotel restaurant isn't bad, but it has a rustled-up look to it, like maybe it's all that was available. A lot of it came out of cans. One doesn't serve Spam for breakfast unless one has to, not in a place that charges as much as a prostitute who speaks a little English. I'm sure there's no special name for the soups of gristle and fat, tasty though they may be. One feels a bit guilty for not finishing what's on one's plate or for helping oneself to an extra half-glass of grapefruit juice when a waiter isn't looking. In fact one feels guilty eating when the waiter *is* looking. Don't they feed these guys? Do they get the leftovers? Or is that what's in the soup?

We UN Experts eat at large circular tables set for eight. When I tell the other Experts I'm going to Kiev to look into Chernobyl, they crack the same did-you-bring-your-lead-lined-jock jokes I had heard several hundred times before I left home. When they finish with that, they ask more soberly what I've done to protect myself and if I fully understand what I'm getting into. I confess I've done nothing and know little. I was living in Brazil when Chernobyl blew up on April 26, 1986. The newspapers there reported the fact briefly, then got back to the comings and goings of the Pope, Princess Diana, Gorbachev, Boy George and whoever else was glittering at the time.

So I don't know what to expect. It wouldn't surprise me to find hospitals full of mutated babies, overflowing cancer wards, forests that glow in the dark, death all around. A Ukrainian I happened to meet in New York heard from his mother in Kiev that all the streets are littered with the flowers that are traditionally tossed at funeral processions. The funerals are constant, she said.

I have two contradictory suspicions. From anyone who claims to know something about the results of Chernobyl, I've heard that the situation is horrifically worse than anyone knows. From the press, I've heard nothing in years. Is it possible that massive slaughter has gone unnoticed or fallen by history's wayside?

By chance I meet a Russian in the hotel who was director of security at Chernobyl. To prove it he shows me his photo identification—with some difficulty, because he's holding a short leash with a Great Dane on the other end, a very cool dog, calm and thoughtful. His master tells me the dog saved seventy-four people after the earthquake in Georgia. I pat the dog's head and say, "Good dog!" Then the director says it is possible for me to take a tour of the Chernobyl plant. Despite the many warnings I've heard, this sounds like a great idea. But then he says I'll have to pay a little money.

It strikes me as rather uncivil to come right out and ask for a bribe from a journalist who's just trying to report on what is arguably the world's biggest problem. I pat his dog again and walk away without a word or a handshake.

Kiev may be dying, but here at the Pribaltiskaya, everything's just fine. I'll give them an A for effort. They've got the marble sink, the shower cap, the little collapsible toothbrush, a fresh disk of soap every morning. The chambermaids don't steal. Nice big color TV, though the only thing on is people talking. The room is never cold: in fact, I have to keep the window open all night. But the towels are threadbare and don't match. God knows this doesn't bother me in the least, having stayed at more than one hotel that has no towels—even one hotel that didn't have beds or other furniture, not even a floor, just dirt, a hotel I'd have to flunk for effort but would highly recommend for its unique and experimental approach to ambiance as it replicated life in sub-Saharan Niger, which is where it was. That hotel offered no collapsible toothbrush. Nor did it have toilet paper, perhaps because it didn't have a toilet. The Pribaltiskaya has all the toilet paper you can use, but that isn't necessarily much because it isn't a paper you'd want to scrape across anything sensitive. It's crepe paper with chunks of wood in it, but it's *there*, which counts for a lot. I carry a folded wad of it in my pocket at all times.

My room on the eleventh floor looks out over the Gulf of Finland, a vast plain of gray ice. Everything out there is gray. The sun does not shine in St. Petersburg, *née* Leningrad, for the five days I'm here. The morning sky doesn't lighten until nearly nine a.m., and by three o'clock in the afternoon it's dark. Most of the time it's raining. Once I shouldered through the windy drizzle to the Gulf just to stand on the ice so I could say I did it. I did it. I stood on the Gulf of Finland about six feet off the coast of Eurasia. Then I went back to the hotel.

A kid named Alex runs an outpost of capitalism in

room 90005. He hangs out near the souvenir shop, passing out his calling card, a scribbled note on a swatch of crepe paper with wood chunks in it. He's got a variety of the usual stuff all over the room: fur hats, matryoshka dolls, lacquer boxes. It so happens I need a leather watchband. He has one with a picture of Yuri Gagarin, first man in space. One dollar. That's within my budget, but a little later I see the price written on the back: a ruble and change, a bit more than a penny.

I ask if he can get me a bottle of vodka, local stuff, not the Smirnoff or lime-flavored Stoly that's for sale in the shop downstairs. He dispatches a cohort on a long shot, but before long he returns with nothing. There is no vodka available anywhere.

I show interest in a mink hat, but forty bucks is too much for me. Alex asks if I've got anything to trade — black shoes perhaps. He's just got himself a dark suit and needs some black shoes to go with it. It just so happens I have a pair of Thom McCanns I don't need half as bad as I need a mink hat. I come back with them and find the room full of Junior Achievers from Iowa. What a confluence of capitalism! But one of them is admiring my mink hat. Quick to take advantage of implicit competition, Alex complains about scuff marks on my shoes, wants ten dollars cash. I give him a quick lesson in the psychological value of cash-in-hand, hitting him with a fin and walking out with a hat that has a story behind it. So what if I have to attend a UN conference in hiking boots? If St. Petersburg can't live with that, they're never going to make the trek to the paradise of free enterprise. A dude in a suit and hiking boots is just the beginning. They ain't seen nothing yet.

* *

My room has a television set, though there isn't much on, except all of a sudden one day there's Mikhail Gorbachev himself explaining something. This is no mere sound bite. For over an hour he talks at a table crowded with officials, reporters, photographers and tape recorders. It's the first time I've seen him up close, not in a black-and-white newspaper photo. I never knew that the mark on his forehead was so blood-red. I can't understand what he's saying, but he seems to be giving a long, thoughtful, personal reflection on something. He talks like a man who's had a bit to drink and has waxed loquacious about something that has saddened him. As he pauses to find a precise word, his eyes examine invisible things in the air.

What he's explaining, it turns out, is the dissolution of the Soviet Union. The country I entered three days ago no longer exists. The place I'm going to — Ukraine — is not what it was yesterday. Depending on who you ask, it's part of a Soviet commonwealth, or an independent country or republic or state, or something as yet undetermined. This situation contributes to the ruckus raised when I alter the UN's plans for my return to the United States. The local people in charge of the conference didn't know I was going to Kiev to look into the Chernobyl situation. My visa is for five days only, and getting an extension isn't easy. You need a written invitation from whatever government agency is hosting your stay in the Soviet Union. No one is quite sure what to do if the Soviet Union doesn't exist, let alone if the traveler has to take a train across a state or republic or country that may or may not be called Byelorussia or Belarus or Byelorus and thence into what may or may not be an independent place that may or may not be called Ukraine.

My best bet for a visa seems to be a young Portuguese woman, Ana Isabel da Silva, who is studying journalism at the University of Leningrad (or St. Petersburg, depending on whom you ask) and stringing for a radio station in Lisbon. She says she can take care of it. She's married to a Lebanese engineering student. She and I converse in Portuguese; she talks to her husband in Russian; he talks to me in a halting but comprehensible combination of Portuguese, French and English. These people know the ropes, one of which is a certain (former) Soviet Deputy at the Pribaltiskaya who has told Ana Isabel that he can get me a visa, no sweat. The deal is that I'm supposed to write and place some kind of advertising for him, something to do with the Free Enterprise Zone or some business he's setting up. But we can't find him. I don't know how that leaves things. I'm not even sure what country I'm in. We decide that, since my visa will expire tomorrow anyway, Ana might as well keep it and try to get an extension. She knows someone at the university who can take care of it. She will mail it to me as soon as I have an address in Kiev. I believe this to the extent that it seems a better bet than wandering around with an expired visa from an expired country.

As for traveling in the evil empire (is it still called that, I wonder, or has that name changed, too?), it is not a problem. Ana says that if anyone asks me where I'm from, I should say "Pribaltico," which may or may not be an independent state or group of independent states, but in either case is full of people who cannot or will not speak Russian. "Pribaltico" is the Portuguese word for that place — the Baltics — but since Russians don't know how to say it, Portuguese is good enough. I believe this the way Ana will procure a visa and get it to me in Kiev. I also suspect that "Pribaltiskaya" is the Russian for "Pribaltico." I will play it by ear as necessity dictates. For all I know, "Pribaltiskaya" means "Baltic Avenue," the square on the Monopoly board between Community Chest and the Income Tax.

For some blessed reason, the Intourist office in the hotel gets me a regular (i.e. not tourist) ticket on the train to Kiev, so it costs $1.36 instead of $78.00. The Russian girl in charge of the UN group, a blonde and bilingual beauty, is panic-stricken to hear of my plans. I must not do it. I do not have permission. I cannot arrive in Kiev without visa or government host. I will be robbed and killed within ten feet of the Kiev train station.

I tell her I'm going to do it anyway. She arranges transportation to the train station. Off I go. A nice driver named Slava gets me and my two incredibly heavy suitcases to the right platform, then, after an announcement from the loudspeaker, over to another platform. He looks at my ticket, explains which wagon and compartment I'm assigned to, wiggles the coupon that's stapled to the ticket and explains that the coupon plus five rubles — about a nickel — will get me "everything to prepare for sleep."

One good sign: four cook-types come down the platform with a cart stacked with slabs of meat and gunny sacks of potatoes. They push the cart slowly, as if to give everyone time for a good look. Heads turn as the cart goes by.

The long, khaki train arrives pretty much on time, whistling, steaming, huffing and squealing as a good train ought to. Slava carries my stuff to the right compartment, heaves it up into a storage area above the door. I slip him a dollar and he seems happy with it.

My mates for the twenty-three-hour trip are two young Russian men in threadbare suits and ties. The compartment is warm and clean and more comfortable than New York's Metro North commuter line and homier by far than Amtrak. The bunks are leather, the walls simulated wood. The window would slide up and down if it weren't cracked and patched over with plastic. The curtain across the window is a hemmed rectangle of linen that has been recently ironed flat except for the ridgelines of wrinkles along the borders. *A cracked window behind an ironed linen curtain.* I have twenty-three hours to think about this.

My second-class mates have not yet looked up from their newspapers. I guess they've made this trip many times. They probably aren't even aware that they're on a post-Communist train headed down through a few hundred miles of radioactive Belarus into the semi-sovereign whatever of Ukraine. Oh, I suppose they *know* that, but they aren't *aware* of it, not the way I so intensely am.

Into the compartment steps the *spitting* image of Leonid Brezhnev, right down to the beady, fat-rimmed eyes, the wrinkled lip of cold command, the mole on the left cheek. Pinned to the lapel of his gone-to-seed gray suit is a gold-on-blue hammer-and-sickle affair which I thought no one would ever wear again. But I've heard the old Communists are making a comeback. Their scant popularity had pretty much dried up after the coup attempt four months ago. Now, they are claiming that the impending famine is an inevitable result of the free-market chaos brought on by lack of government control. Under the Communists, the lines were shorter. Apparently the old people remember it that way. Having dedicated their lives to trying to make it work, having risen from the frozen swamp-mud of Czarist feudalism, they don't like seeing Communism abandoned. And some of the young people believe what they hear about the good old days of Stalin.

Many are the birch in Belarus. Belarus is all set for birch. Also lots of evergreen and clumps of brown grass under swatches of snow. The houses are nothing fancy, just plain old down-east contraptions sheathed in dark shingles. Everything isn't as dead or uninhabited as I'd expected. Nothing glows with radioactivity. I saw some kids playing hockey on a rectangular pond that drains the raised railbed. I haven't seen a motor vehicle since we left St. Petersburg, but it looks like everybody's got electricity and a TV antenna. I've certainly lived in worse places than the places we're passing.

Old Brezhnev has brought a nice little lunch with him. He ambles down to the end of the car to draw a mug of hot water from an explosive-looking brass samovar. After returning and sliding the door shut, he lays out lunch on the table that juts out a foot and a half from the window. He cuts a thick slice of bread from a dark half-loaf in a plastic bag. He's got a can of sausages with a gaudy label in Russian and Chinese. He puts a pinch of black tea into the mug, then pours half a glass of vodka from a half-liter bottle. He's also got two hard-boiled eggs. He downs the three or four ounces of vodka in a single gulp, chases it with sausage-water from the can. He taps a timid crack in the shell of one egg, peels it as quietly as he can. For the first time in my life, I become aware that the peeling of a hard-boiled egg has its own identifiable sound.

Our younger guys on the top bunks look as if they haven't eaten in a few days. Their Goodwill pants, sport-coats and frayed (but unloosened) ties hang on them with full droop. I pass them each a Halls cherry cough drop (flown in from Lebanon and given to me by Ana Isabel) and they look pretty happy. One says "Thank you" in plain English. I can't imagine how he knows I'm foreign. I haven't spoken a word and I'm wearing a brown tie with fleurs-de-lis on it and a green plaid shirt and tweedy jacket with leather patches on the elbows. I thought I looked pretty much like the average Russian beezneezman. Maybe it's the glasses. I'm beginning to notice that no one else wears them.

I can tell Leonid feels a tad guilty about all his food, especially the vodka. He pours it out (his fourth Texas-size shot now) while I seem not to be looking and dumps it down his throat. I guess he'll be asleep soon.

Now for a stab at some food myself. The "PECTO-PAH BArOH" — which, once you sound out the Cyrillic letters, comes phonetically close to "Restaurant Wagon" — could pass for a miniature diner in Montana or some place like that, except that they keep the loaves of bread under the bench seats. The table top is chipped Formica. A real flower, a zinnia I believe, stands alone in a slender green vase of crude glass. Napkins — rectangles of newsprint, actually — stand in a fluted plastic cup, a plastic reminder of Greek grandeur. Lots of cups of juice stand on two stacked trays, but I'm not going to drink any. It's probably St. Petersburg tap water, which has visible entities floating around in it. If you look down through a glass of it, you can see a brown tint. If you let it sit long enough, it settles to a sedimentary film of mud.

As for the food, I guess I and half a dozen other people are waiting for it. I have come to understand it becomes available at three o'clock. I assume it just arrives in dishes, unspecified, unbidden. Sitting there means you want what they've got, the blue-plate take-or-leave-it special, even if it comes in an aluminum dish. The pleasant smell of cooking meat comes from the little kitchen where a standard-issue 250-pound cafeteria lady squawks orders. It's just after three and I haven't eaten anything since breakfast except for a Hall's cherry cough drop. It's probably the longest I've gone without food in over a decade.

Well, it seems you can and indeed *must* order your meal. A waitress built like an IL-86 hits me with a bunch of syllables that all seem to rhyme with Zhivago. I say, "Yohnny Pawnny Maio," my version of "I don't understand" in Russian, clear as a bell. She says Zhivago a few more times and I say, "Da," hoping it's not for liver or beets. She writes it down. It turns out I've ordered a quite-edible plate of gristly beef chunks, potatoes and beets. I down all but the beets. This is a psychological problem. Apparently when I was a little boy in a high chair I smeared my face and shirt with beet juice while my mother was out of the room. When she came in, she thought it was blood. She cut loose with a scream that still echoes in my subconscious, preventing me from stomaching a beet. This doesn't bother me a bit because I hate beets.

Back in the compartment, I finally strike up a conversation with old Leonid. I use the term "conversation" loosely. He knows no English at all. I know half a dozen words in Russian. "Zhournaleest," "Chernobyl" and "Amerikanskiy" are three of them. It's enough. He knows why I am here. He gestures toward me with a glass of vodka. I decline with a string of *nyets* and humble body language. But he insists. I hold the glass and look at the stuff, take a whiff. He gestures to tell me to dump it all down at once. I whistle at the challenge, then toss it back. Wow. Nice. Yes indeed. He laughs, his eyes disappearing in the wrinkles of his face. It isn't the first time I've played this game. I've played it in Mexico, Peru, Brazil, and more places than I can remember. All I remember is that I always lose. Leonid pours more. I turn the tables, making the gesture that means down it all at once. He does it. We're halfway through the bottle. He pours more, I decline. He leaves the glass on the table and asks something I don't understand. He knows that I can't understand him but is unable to resist talking. I know this scenario. I've run into it in more than one bar. He's drunk and I'm a captive audience.

One of the younger men takes pity on me and translates. Brezhnev wants to know what paper I work for. I tell him I'm writing a book. The younger man asks me why. I tell him it's because no one remembers, or even knows, what happened at Chernobyl and what's happened since. So we talk about Chernobyl and other things for a while. The young guys, it turns out, are brothers. They've been touring Europe, selling art from their gallery in Kiev. I ask about the Communist symbol on Leonid's shirt. They say there is no doubt but that he is a Stalinist and proud of it. But he is a generous Stalinist; everyone gets a dose of his vodka. Then we retire to the little compartment between the cars for a smoke. In the haze and rumble we crack jokes. I tell the old man that on trains, all the good people are between the cars. He laughs himself into a coughing fit.

CHAPTER TWO

Wormwood

How to explain fate? I get off the train into the rush and press of the station at the close of day. The two directors of the Kiev art gallery help me lug my leaden suitcases down the platform, down some stairs, down more stairs, and across a crowded lobby to a taxi stand. A dozen drivers crowd around looking at the handwritten address of Ljudmula Mikitichna, a person who speaks English and has agreed to be my translator — my only contact this side of the Mason-Dixon Line. I had transcribed her name and address from the original that was written by a helpful Ukrainian who lives in Connecticut, so the letters aren't just right. But when the drivers all start chanting the same word together, I know they've got it. Off I go with a certain Volodya in his rattly little sky-blue Lada. It isn't a taxi; it's just his car, a Soviet imitation of a Fiat. As we drive down the dark and mysterious streets of Kiev, I note that like every other city in the world, this one has its own peculiar smell. Here it's a nauseating essence of rotten eggs frying in cheap vegetable oil.

Volodya is a big, ugly galoot with the biggest gut in town, a hairy mole on his chin, and a gold tooth bent over one canine. We hack out a crude, rudimentary language consisting of half a dozen words in English and about as many in Russian. I manage to tell him I'm a journalist working on a book about Chernobyl. By holding one fist to his eye and cranking another around his ear, he tells me he knows a filmmaker who just made a film about Chernobyl.

After a half-hour drive into a suburb of ten-story housing projects, Volodya turns up on to a sidewalk and follows it among buildings. How did he ever know this address? Or is this where he's taking me to kill me? But he drops me off at the right place, even checks to make sure the door to the building, which looks like a back door to me, is open. He asks for ten dollars, which I know is absurdly expensive. I offer rubles but he wants dollars. Okay then, five dollars — still way too much, but it was a long way indeed and who needs problems in a place like this?

I lug my suitcases up three floors and leave them behind as I walk up to the sixth. It's a dark, narrow concrete stairway. It reminds me of the stairway Winston Smith climbs in the first pages of *1984*. As I recall, at about the fourth floor I'm supposed to bend down to massage a varicose vein in my calf. Why do I remember such a thing? I don't have varicose veins, but of course I just got here.

At the top of the stairway are three doors. They are padded, dimpled with large buttons which hold the padding in. I can't read the numbers on the doors so I choose one and thump on it. It opens partway and a kid and a lady with red hair lean out. All I can say is "Ljudmula, Ljudmula" and show her the wrinkled paper with the address. The lady points to the door directly across the landing. I thump on it. An old, old woman opens the door. I say "Ljudmula" a few times and hand her the paper. She talks at me. I say "Yohnny Pawnny Maio. Ya Americanskiy." She talks more. Finally we arrive at "*Ljudmula nyet.*"

Now I'm stuck, really stuck. I don't know if this means Ljudmula isn't home or doesn't live here or what. So what am I supposed to do? Even if I had somewhere else to go, there are no taxis in this part of town. We hadn't seen a car for the last ten minutes on the way in. I doubt I could even find the road, let alone carry my suitcases that far. It's twenty degrees outside and very dark.

But lo! Volodya returns! He found a document, a receipt for some money I changed, on his back seat. I say, "*Ljudmula nyet.*" With our dozen common words we construct the common understanding that I'm in trouble. He says "*Bandido*" and draws a finger across his throat. Now what? He goes upstairs and talks to the old lady but then can't explain to me what she said. Gesturing, he says, "Led's go."

He takes me to his apartment, a two-room place some forty minutes away. He puts out a spread of canned meat, sausage, cheese, buttered bread and pickled garlic gherkins and vodka on the little coffee table. We watch American soft-porn on TV and call Ljudmula a few times, but nobody answers.

So we go see his filmmaker friend, Andrew Loskutov. Andrew speaks lots of words in English but without grammatical foundation. He tosses them out in random order, tense and person, stitched together with random prepositions. I understand perfectly. He's a very nice guy. Same goes for his wife, Irena, and their nine-year-old boy. Another couple shows up and we talk about America, my impressions of Kiev and St. Petersburg, and Chernobyl. They have a few horror stories. They've heard about five-legged dogs and two-headed calves and such, but they're rare as hen's teeth because the government, in its continuing effort to keep the problem a secret, destroys such evidence. Andrew tells me about a tree that grew near there, a pine with two branches that grow straight out like a cross. There's a little shrine around it and some graves. The area is supposedly holy. During World War II the Germans used to hang people from the branches. But right after the explosion, a storm blew the tree over. There were plans to save it, but then the radiation killed it and the government came and sawed it up into little pieces and buried them, as they did with anything else showing signs of radiation.

Andrew's wife, Irena, asks me if I believe in God. I allow as yes. She tells me Chernobyl was predicted in the Bible. Yeah, right, I think; so was everything else. But Andrew pulls out a Russian Bible and an English Bible. He finds the relevant lines at chapter ten, line eight, in the Russian version, but we can't figure out which book it is in, so we can find it in the English version. I unscramble the letters from the Cyrillic, but it doesn't sound like anything. He gets a Russian-English dictionary and looks it up. It says, "Frankly." We paw through the English Bible. The Gospel according to Frank? Not likely. Proverbs seems as close as anything — proverbs are frank expressions, right? But Proverbs 8:10 says *Receive thy instruction, and not money: choose knowledge rather than gold. For Wisdom is better than all the most precious things; and whatsoever may be desired cannot be compared to it.*

Wise words indeed, and frank, too. But that's not what we're looking for. What we're looking for, I finally figure out, is a prediction, which, of course, is in "Revelations." We paw through to the right place. Yes, this is it: *And the third angel sounded the trumpet, and a great star fell from heaven, burning as it were a torch, and it fell on the third part of the rivers, and upon the fountains of water: And the name of the star is called Wormwood. And the third part of the waters became wormwood; and many men died of the waters because they were made bitter.*

Wormwood, says Andrew, just happens to be an ancient Russian word for a certain plant. He gets an encyclopedia of plants. We find the right one. In Latin, it's *Artemisia absinthe*. The English word for that plant, he says, is Wormwood. The word for Artemisia absinthe, in ancient Russian, is *Chernobyl*. Chernobyl's right there in the Bible. It's downright apocalyptic. How to explain fate? I haven't been in town three hours and my finger's already on the subject of my book. It's right there in the Bible.

Wormwood.

* *

Tuesday, December 17, 1991

Trying to get organized today. Trying to make progress. I know how things get bogged down if you don't keep them moving. I don't want to waste any time here, but I'm hitting obstacles everywhere. From Volodya's apartment, where I spent the night, I call Ljudmula again and again and again, but she never answers. I guess she's still away. Volodya says and Andrew translates that it was her apartment we went to last night, and it was her mother who answered the door. Her mother did not seem to me to be the sort of person who is at ease dealing with a telephone, so I assume Ljudmula is still "away," which is what her mother told Volodya and Volodya told Andrew and Andrew told me.

I also have the address of one Ida, mother of the Ukrainian in Connecticut who gave me Ljudmula's name and address. But Ida seems to have a phone chronically off the hook. I have the sensation that all of Kiev is chronically off the hook — busy but not getting anything done, at least not in terms of my needs.

First thing in the morning, Volodya takes me back to Andrew's and thence to the film studio to see Andrew's film. But we run low on gas — natural gas, it turns out. Volodya keeps half a megaton in a horizontal red tank in his trunk. It's a handy alternative when you can't find gasoline. But of course it means you have to find natural gas. Volodya takes us to a place deep in an old residential neighborhood of brick houses: at the back door of one of them a black market natural gas business is in operation. Volodya brings another red tank out to the car, rolling it along on its bottom rim. As he hooks a hose to the tank in the trunk, gas leaks all over the place. What a stink! Now I know why Kiev smells the way it does. It's not eggs gone bad. It's Volodya's car. Andrew walks down the street for a smoke. I walk up the other way, just in case. Refueling turns out to be a complicated operation involving hot water poured over the source tank from a tea kettle. The steam makes for a good photograph.

We don't drive far before the engine has an asthma attack. We pull over to the shoulder of a wet, gray freeway. Wet, gray trucks swish by. The temperature's just above freezing, or maybe just below. In either case, it feels colder because I'm in my tweed jacket, not my coat. I pull a sweater out of my suitcase while Volodya monkeys with the regulator that the natural gas feeds through. He has to switch over to gasoline to get the engine running again.

Volodya drops Andrew and me at the Ukrainian Environmental Protection Society, a single dark hallway with a dozen offices and a bathroom that smells far worse than some outhouses I've been in. Andrew has a small office there. It doesn't look permanent, but there are a couple of wooden desks and a phone—a party line that rattles every few minutes, never for him. It looks like the kind of office that maybe a dozen people use, but only now and then, none of them really considering it more than a place to sit down for a smoke before moving on.

Part of this government agency is what they call a hotel. It's just a couple of rooms with three beds each, where people visiting from other cities can crash for a night or two when they visit on environmental business. As I understand it, there are no hotels in Kiev except those catering to foreign tourists. They are as expensive as American hotels, a single night costing more than the average Ukrainian makes in a year. Well, I'm here on environmental business, so the big boss agrees that I can stay here for a couple of days until a certain conference begins. Then I must leave for a couple of days, and then I can come back. It wouldn't hurt, says Andrew, if I offered said boss a tip of twenty bucks. I move into Room Sixteen. My stuff sprawls over all three beds, the little desk, the closet, the credenza, the window sill, even outside the window sill. Home sweet home.

We go to see Andrew's film at the studio, which is just up the hill from the Society. The film's okay, but I wish I could have written the script, or even just translated it, and done some of the camera work. I could tell they were hard pressed for appropriate images. They arrived at the scene several months after the tragedy. Empty houses, a headless teddy bear on a cracked sidewalk, weeds breaking up through asphalt, laboratory mice with tumors the size of golf balls hanging from their bellies. The mice have been living near Chernobyl since the accident. Their lifespans are thirty percent shorter than normal. The lab people hold them up with large tweezers. In an emotional voice the narrator says that the town of Chernobyl was founded a thousand years ago and everything was just fine there until they built a nuclear power plant. Now the richest farmland in the Soviet Union has become deadly instead of fruitful. The film says a million people died, but that seems like an exaggeration to me, unless you consider that maybe a million people will have a lifespan thirty percent shorter than normal.

About half the film was a quasi-incongruous report on a California-style international protest demonstration held just outside the thirty-kilometer "Prohibited Zone" that isolates the plant. The event was called Next Stop: New Planet. Everybody took off his/her clothes in protest, painted images of leaves on their bare skin to symbolize nature and to show how little protected the body is against radiation. There was some skinny-dipping in a radioactive stream. The camera grabbed some quick T&A shots, all of girls, none of boys. Must have been quite an event.

Andrew says I can buy a copy of the film for a hundred dollars. It would be useful to me and would also be doing him a very big favor. He'll let me have full rights to do whatever I wish with it in the US. That sounds like a good deal to me. It's a 35mm film, so it will be easy to shoot still pictures from. I'm most interested in the rats with the great big tumors. He also gives me a video version of the film.

We eat in the institute restaurant, a rugged little meal of pork burgers (called *cutlets*), noodles, shredded cabbage, blinis and an unidentifiable juice. The room has a high ceiling and decorative plaster patterns up the walls.

Tall windows are covered with patterned curtains. The chandelier hints at better times back when cutlets were cut, not pattied.

For lack of choice I've been drinking tap water. It looks cleaner than St. Petersburg water, but something in there is having a heyday in my bowels, a gala duodenal soiree. I can hear the merry microbes whooping it up in there. I beg off Andrew's attempts at entertainment and go to lie down for a while. I stop in at the bathroom.

The toilet's in a concrete stall. It seems to be a regular toilet but with a brick platform built around it up to the rim. You walk up a step, do an about-face, put a boot on each side of the rim, and squat down to do what must be done, holding on to the door handle to keep from slipping and dying an ugly death.

The hour's sleep does me good. I take a diarrhea pill, wash it down with water I find in a clear glass pitcher that comes with the room. Then Volodya arrives in a blossom of natural gas. Andrew is with him. We go off to find Ljudmula.

Glory be, she's home. Her phone, it turns out, doesn't ring. It only goes *fltfltfltfltflt,* which she can't hear. She was wondering what had become of me. She wasn't expecting me until the next week so her apartment, which she was preparing for my visit, is in the midst of a total overhaul, including a complete repapering of the walls. The furniture is stacked all over itself. She's very apologetic. In fact, she won't stop talking. Her English is sweetly British, virtually perfect, a tad archaic. Tall and poised, lean and maybe slightly past sixty, she has a rather regal bearing.

Ljudmula has already lined up various people for me to see, some of them pretty important government people. She has also lined up a few places for me to live. The first is with two elderly sisters who live downstairs in her building, Rachel and something that starts with a G. We go visit them and check the place out. It's a rather spacious and very clean apartment. They show me my own room. Very nice indeed. I'm afraid now of offending Andrew, who has offered me his apartment for the two days of the conference when the hotel will be closed. But Ljudmula won't take no for an answer. In fact she won't take any answers. She's not the type to stop talking for the sake of a breath or a bit of data. So we agree I will move in with Rachel and G— for two days. I don't expect it to be too bad. They look extremely sweet and motherly. Their smooth faces are lined with tight wrinkles and peachfuzz and they are smiling, smiling, smiling. Rachel rubs her hand up and down my back every time I'm within reach. She's in love and so am I.

Ljudmula is worried about the company I've fallen in with. She finds Andrew's English atrocious and his Russian lowbrow. Without seeing Volodya — he is staying outside to guard his car — she doesn't like the looks of him, either. She feels responsible for me and doesn't want to see me murdered while I'm under her wing. I don't want to see that either, but it doesn't seem like a problem to me.

On the way back to the hotel we stop for mineral water. I discover Belgian beer can be bought for nineteen rubles a can. Andrew tells me it's too expensive, but I tell him it's not. I buy four cans and also five bottles of what they call lemonade. Then we drive a couple of blocks to a place that might have mineral water, but I tell them I can get by on beer and lemonade.

Back at the hotel the beer really cheers me up in belly and mind. I drop another pill, drink some lemonade. It bears no resemblance to any lemons I've ever tasted, but there's a hint of citrus and a wallop of sugar. Maybe it isn't tap water and maybe it isn't radioactive and maybe it'll do me more good than harm.

* *

Morning arrives and with it, Volodya and Andrew. They have new plans for me. Volodya's short on gasoline — he needs at least a little to get the car running in the morning — and doesn't want to drive to Rachel's. Better, he says, if I stay at his apartment and he goes to live with his brother. It doesn't quite make sense to me because we still have to go pick up Ljudmula to translate for me. But I don't have the language to argue out the details. I agree and off we go to Volodya's apartment. He installs me, then goes off with a bottle of homemade vodka ("*samagunke*") to barter for some "benzene."

Content to hang out for a while, I boil some water and brew a big mug of the Earl Grey tea I brought. I wash my clothes in the bathtub, a deep, claw-footed thing with no plug. Then I hang everything on the little enclosed porch, a space the width and length of a large bed and stacked deep in old tires and car parts.

Let the Western World know that I went out and bought bread and it took me only five minutes, and it took that long only because I walked the long way around the building to the state store that's on the first floor of the apartment building. Granted, there wasn't much bread to buy, and not much else either. I took four fist-sized loaves. Other stuff available seemed to include jars of boiled or maybe pickled beets, strawberry jam, and yellowish juice in gallon jars. I bought a small jar (about a pound) of the beets, thinking it was cabbage. Total cost for that and the bread: 1.4 cents. Total time in line: sixty seconds max, and would have been less if I'd been a little snappier with my kopecks.

Volodya comes home for lunch. We eat soup made of boiled beef by-products and noodles, raw garlic cloves dipped in salt, slices of something like kielbasa. Volodya takes his soup with the hearty slurp of the streetsmart. Somehow we converse. Among the things I learn is that he is my age — forty — and has been an orphan all his life.

Ljudmula calls to say she hasn't quite lined up interviews but soon will. So Volodya and I go out in search of a barbershop and a fur hat. The first barbershop has a few people waiting so we go to a store that sells things for hard currency only. Wares include electric mixers, a plastic patio table, a stove top, a few whiskeys, boots, jackets, a vacuum cleaner, but no hats.

We take a thrilling ride down a lot of odd streets,

some just muddy trails down hillsides of ramshackle houses. Everything is gray with drizzle, sleet, snow, soot and the slate-dark light of an overcast sky at a latitude where the sun at noon is only a few inches above the horizon, or so I suppose. I've seen it only once in the past ten days, and that was in St. Petersburg, far north of here.

If Volodya drove like this in a civilized country, he'd be locked up for attempted suicide. He hovers within three inches of other cars going sixty miles an hour down the wrong side of a six-lane boulevard. He passes even when another car is coming, climbs up onto a sidewalk to get around a bad puddle, whips the wrong way down high-speed one-way streets. If this were my mother driving, I'd be worried. But Volodya has been driving underground taxis for ten years. The kamikaze of Kiev is a man at one with his automobile and the ragtag flux through which it flows.

I live long enough to get my hair cut by a middle-aged woman with a husky voice, and tender, careful hands. She trims my beard as well. She chats with Volodya. Their discussion involves a lot of numbers which, being about the only words I know in Russian, glare like fireworks in the black sky of their strange language. I strongly suspect he's telling her to charge me a lot because I'm a foreigner. When she's done, she says twelve rubles, which is certainly affordable, though I hate to think I'm being ripped off. I give her multiple *spasibas* and a ten-ruble tip for her nice gold-toothed smile. As I hand it to her, I'm *sure* Volodya says, *"See?"* He and she and a couple of other people burst out laughing.

* *

It's amazing how Volodya and I manage to communicate. We share a total vocabulary of under twenty words, but somehow we exchange a lot of information. We have, in effect, created our own language. We have gestures to indicate "tomorrow" (a forward chop of the hand), "o'clock" (a twirl of the finger at the wrist), and "lunch" (pointing down the throat). Neither of us knows the days of the week in the other's language, but we know what they *sound* like, so we make that sound and tick off our fingers — "hmmmday, hmmmday, hmmmday . . . " until we get to the right day. He doesn't know, off the top of his head, the word for, say, the number nine, but can start counting at one and keep counting until he gets there. At his mechanic's garage he uses the international gesture that means "The rear wheel bearings are shot and it's going to cost a fortune to get new ones."

We go to Volodya's girlfriend's apartment. She's wearing a white terry cloth robe worn smooth with time. She doesn't look too pleased to see him, or maybe me. Her name is Natasha, which strikes me as somehow appropriate for a girlfriend-on-the-sly. (Volodya is married, but his wife lives in another city.) Natasha is a high school music teacher. In her tiny kitchen we sit down to lemonade, cheese and tea. Little by little I figure out Volodya is staying here, not at his brother's house. I notice that Natasha's face tends to settle into a look of peevishness but readily bursts into a smile when Volodya cracks a funny. He's a clown. His stories seem to be of the sort that end with him holding out the palms of his hands and saying something like, "So what was I supposed to do?" Natasha laughs every time. Even I laugh. We make a lot of jokes about *samagunke* vodka. She has a jar of it fermenting under the table. They try to explain how to make it, but we lack some key vocabulary. It takes flour, sugar, water, yeast and something else. Their drawn-in-the-air diagrams seem to indicate mushrooms.

During this odd talk I learn that Gorbachev has finally fallen. Volodya mentions it as if in afterthought: "*Gorbachev kaput.*" Natasha laughs, and Volodya gives a quick whistle as he points his thumb back over his shoulder. "*Kaput,*" he says. "*Gorbachev goot-bye.*"

Then off to Andrew's house. We drink his harsh *samagunke* and laugh a lot. He shows me the device he uses to distill vodka from a pressure cooker. It's a clear plastic tub about fifteen inches long with a ripped clear plastic tube inside. Three rubber hoses dangle from it. A beer can is taped underneath as a handle. The tube tapers at the end where the condensed alcohol drips out. It looks like a bizarre and dangerous marital aid. I want to take a picture of Andrew holding it, but he says "It is not interesting for a film director to have such a thing." He laughs up phlegm from his chest. Volodya, however, is proud to pose. He hasn't had a drink in four years. He holds out his hands to show how they used to quiver.

CHAPTER THREE

Pravda

Volodya drives Andrew and me to *Kiev Pravda* (*Pravda* means "Truth," an ironic name given to many papers during Soviet times). We're supposed to meet Ljudmula, who is to introduce me to a certain journalist who wrote a lot about Chernobyl. But while we're there, Andrew meets a fellow English student. She takes us around to meet various people. When we get to the editor-in-chief's office, he happens to be talking to the fellow Ljudmula was to introduce us to, Grigorio Dmitrivich.

While we are upstairs Ljudmula has arrived and, unbeknownst to us, is waiting for us in the lobby, where there's a kind of wake going on. A woman who had just been laid off got so upset that she died. Ljudmula, no spring chicken, has to hang around looking at this lady's picture and smelling her flowers until Volodya goes to her and asks if perchance she is Ljudmula.

Grigori Dmitrivich was at Chernobyl right after

the accident when "liquidators" were trying to push a steel frame up to the reactor and fill it with cement, the beginning of the sarcophagus. Other liquidators — workers who had built the Kiev subway — were digging a tunnel under the reactor to reinforce the floor. It threatened to collapse under the weight of all the sand dropped atop it by helicopters. Everyone was accumulating very dangerous doses of radioactivity: Grigori ruined his lungs by being there for just a few days. Now he suffers terrible asthma. In 1989, while hospitalized, he met an old friend who had once been a strappingly healthy man. Working as one of the miners who dug the tunnel under the reactor, the man had taken a dose of 140 rad. (Five rad is the maximum allowable annual dose for a nuclear worker in the United States.) Three years after the accident he was shriveled up and too weak to hold a spoon. He had a constant fever of 38 degrees Celsius (100.4° F). His official ailment: sore throat.

Grigori says that no one knows much of the truth about Chernobyl because it happened in 1986, when such things were still kept secret. Now the Soviet government has pulled out. Whatever information it had, it took. What little information exists in the public is scattered among doctors, physicists, meteorologists, the departments of environment, energy, health, the military. No one knows who else has a piece of the puzzle.

The Soviet government had promised a lot of aid and compensation to the liquidators who cleaned up some of the mess, and to people living in contaminated areas.

But they delivered too little to do anyone any good — everyone calls it coffin money — and now the Soviets are gone. The Ukrainian government has made similar promises, but it's too broke to deliver. Grigori wants me to thank the countries that have sent aid, but he says that he personally hasn't seen any of it. He suspects it was gobbled up by the bureaucrats.

Grigori is angry about the accident but not angry enough for Andrew, who later calls his position "official." Apparently that word means "offering something less than the full truth." Andrew explains that the reporter would lean toward the official because if his newspaper pisses off the government too much, the company might be denied such state-provided supplies as paper and ink. The Communist bureaucracy is still strong. It can still mangle any wretch so foolish as to fall into its gears. He asks me not to thank the countries that have sent aid but rather to tell them to stop. They are only feeding the bureaucrats and making them stronger.

* *

To my utter astonishment, the next day is the first day of the International Conference on Chernobyl right here in beautiful, historic Kiev. Why didn't anybody tell me this? It's being held by *Soyuz Chornobyl*, the Chernobyl Union, which is composed of liquidators, refugees and anyone who is conceivably a victim of Chernobyl. Over the course of several phone calls I schedule Ljudmula and

Volodya to get me and her there. Ljudmula brings Elena, a twenty-seven-year-old plant biologist. Elena's rather uptight and self-conscious but quite quick of mind. Her laugh is forced and artificial, sliding out the side of her mouth when some conference speaker says something worthy of cynical response. I suspect her of poetic tendencies because she's quick to see metaphorical parallels. When a conference speaker reports that deaths in Kiev now exceed live births, she notes that Ukrainians are an endangered species. When they say Ukraine must stop being an international beggar and should contribute something to the world, she adds that Ukrainians should share their diseases, some of which are very original. The new collectivism, she says, is the sharing of radioactivity in all their bones. This, she says, is the only fair collectivism she has ever witnessed. Every time she pops up with comments like this I fall a little in love.

The conference starts out pretty well with a troupe of young Ukrainian folklore dancers in full costume, most of which is red. Their group is the Lelechenky, all children evacuated from radioactive areas. They actually look as if they're having *fun* prancing around on stage and singing what in America would be dumb songs like "Casey Jones," or "Sweet Betsy from Pike." American kids don't have fun singing such songs. They do it because they have to. It's as much a part of current culture as penguins and kangaroos. But these Ukrainian kids are wearing costumes and doing dances that are four times as old as the United States. Centuries-old folk songs are part of contemporary culture, and they're fun. What's most fascinating is that between the thirteenth century and December, 1991, Ukraine lived almost constantly under conquerors with different cultures. It was really independent only from 1918 to 1922. Over those seven centuries, Ukraine suffered innumerable slaughters by conquerors from the Mongols to the Tartars to the Germans to the Russians, ending with Chernobyl.

The president of the Union comes on to remind everyone that they are there because of the children. They are the real victims. A million of them still live in contaminated areas. An unknown number have lost a parent who was a liquidator and who died of a sore throat, bronchitis, diarrhea and flu — the diseases that doctors were required to list when the workers died of thyroid problems, lung cancer, blood diseases, gastro-intestinal collapse, hemorrhage and heart attack. Ninety percent of children in contaminated areas say they see no future for themselves. Thyroid ailments have become common among them, increasing by several hundred percent. Aid from UNESCO was lost as it filtered through the bureaucracy in Moscow. Fifty million yen from Japan never came through that filter. A donation of coffee from Brazil went into the filter and came out on the black market. A load of vitamins went into the filter and bureaucrats gobbled down fistfuls of the stuff. A telethon broadcast across the Soviet Union raised millions of rubles for victims but it all went to finance the building of the sarcophagus.

The Union wants Ukraine to be a nuclear-free state. It wants to find out what happened to several tons of nuclear fuel that was in the reactor. Something like fifty of the 230 tons cannot be accounted for. At the time of the accident, the government said not to worry because most of the radioactive material stayed in the reactor. Now that they can't find it, they say not to worry, it was blown into the atmosphere. One theory says that the stuff is still in there. Another is that it was "mined" and used for nuclear weapons. There are 4,000 nuclear bombs in Ukrainian territory. Nobody knows who gets to keep them. Maybe a country that no longer exists; maybe a country that's two weeks old, flat broke and for the time being is the world's third largest nuclear power. They're not even sure who their navy belongs to.

The conference goes downhill after the entertainment and introductory speeches. It turns into a congressional-style meeting bogged down in formality and debate. One woman, an official in the Union, goes to the podium and says, among other things, that she just can't keep going any more. It's too much work, with too little progress and too little thanks. She stops when she weeps so hard she can't raise her head from the podium.

The big issue is whether to turn Children of Chernobyl into an international organization. (I later learned that this Children of Chernobyl has nothing to do with the Children of Chernobyl Relief Fund headquartered in Short Hills, New Jersey, though both are legitimate and benevolent in their projects.)

Fortunately — someday I'm going to pay for all this serendipitous fortune — there's a special trip to two hospitals. Only foreigners get to go. A couple of Germans, a French, a Canadian, a fellow from Hartford, Elena and I and a few Ukrainians get onto a bus and off we go.

The first hospital is for children. Unfortunately we don't get to troop through the ward because the children would feel like animals in a zoo, says the lady in charge. But fortunately there's a nice little spread of food in the director's office. By nice spread I mean a bowl of beat-up apples, some tangerines, and wafers wrapped around a carmelesque filling. Elena shakes her head. "Normal people like me don't often see so much food," she says. She eats two apples and four tangerines — she's a vegetarian — but turns down the wafer roll for fear of getting fat. I slip a couple of apples into my briefcase for her.

While we munch, the hospital director, a woman named Zoroslava, gives a little speech. In a thin and tremulous voice, she parts with some useful information, such as the fact that the apples we're eating are radioactive, as is everything else in town. She says seventy percent of children born in Ukraine have health problems. Twenty percent of families can't bear children. Forty thousand women can't get pregnant. Seventy percent of newborns have health problems. The death rate has doubled and the birth rate has dropped. What good is Ukrainian independence, she asks, if the population is dying off?

The hospital has chosen a certain rural district that has 1.5 curies of cesium per square kilometer, qualifying it as Zone Three, for a two-year study. (There are four zones of contamination. Four is least contaminated.) A handful of doctors go there weekends to monitor the effects of long-term low-dose radiation. Zoroslava says these are people who have been forgotten by God. They've also been forgotten by doctors. Of the twenty-seven doctors who used to live there, all but two have fled. Seventy percent of the children are eating contaminated food. Seventeen percent of them have thyroid abnormalities. The cancer rate has doubled, especially in women's reproductive organs. Everyone's red blood cell count is low. Children are inclined to be ill more often than normal.

The doctors are discovering that long-term low-dose radiation attacks the pancreas and a tract of the liver. The doctors could learn more if they had a portable ultrasound equipment. The world, says Zoroslava, should be dedicating massive effort to this unprecedented opportunity to study the effects of radiation. Instead, the world is ignoring it.

The team is going to this town on Christmas for a couple of days. I ask if I can go. Zoroslava says they would be delighted.

Next we go to a pediatric hospital for a look at the neonatal department. The head doctor reports that they treat 800 premature babies each year, the same number as before the accident though the total number of births is way down. This year they have lost only eleven of the thirty-four babies born under one kilo (2.2 pounds) — a miracle considering they have few diapers, no heating lamps, no IV regulators. They have to resharpen syringes until they're too short to use. Fifty-two percent of babies are born perfectly healthy; the rest have disease or high risk of one. The infection rate is high because adult patients are interned nearby.

Among the visitors is Alex Kuzma, director of the New Jersey-based Children of Chernobyl Relief Fund. He's a Ukrainian-American who, like me, is from Connecticut. It turns out we have a friend in common. He's in Kiev with the mission of setting up a system by which his organization can send aid without it ending up in the wrong hands. In the past, they trusted local agencies and organizations and saw some of their aid end up elsewhere. Now they have reliable contacts and a system that involves paperwork, proofs, back-ups, and so on. It will be harder to steal food from the mouths of irradiated babes.

I get all excited about the trip into the countryside. It will add a whole new dimension to my research. I want to write a book not only about radiation and scandal but also about just plain people and how they live. I want to drink tea in their kitchens and play with their children and scratch behind the ears of their little Ukranian dogs.

But the next day, when I call back to confirm, I'm told that the trip has been canceled. The car broke down and no parts can be had until March of next year. I'll have to wait till then, and so will the research project, and of course so will all the sick people.

CHAPTER FOUR

Zones

Volodya has to go see his wife and son for the weekend. They live seventy kilometers from Kiev. I'm to stay in his apartment until he gets back, then move back to the Environmental Protection Society's hotel. Volodya's friend (who has the same name; I think of him as Volodya II) will take responsibility for me. I'd say he speaks about twice as many words in English as Volodya I. That puts him up around the range of a couple dozen. Also his car is newer and closer to being comfortable.

He picks me up at Volodya's house at one o'clock so that we can meet Ljudmula somewhere at two. But, as it turns out, there is a delay in plans. We arrive at an unidentifiable stretch of sidewalk fifteen minutes early, and sit in the car looking at the drizzle and snow until a big guy in a flattened fedora shows up with a limp flower. He hangs around on the sidewalk for a quarter of an hour, then finally comes over and asks, in crisp, perfect, forced English, if I'm the guy he's looking for. It turns out I am. Well, Ljudmula's going to be an hour late. Well, okay, now what? Well, now nothing. We wait. I sit with the door half opened while this poor guy leans vaguely down, afraid to pull away, afraid to get in. I warn him that his tan trench-coat is up against the muddy car. He says that's not a problem. But after ten or fifteen minutes he says we should go to so-and-so's house and Ljudmula will meet us there. He will wait for her. He hands me the limp zinnia and says, "Please, give this to the hostess. Hostess, yes? Hostess."

So Volodya II takes me and the flower to a nearby address and up the rattletrap elevator to the cramped, clean apartment of a chemist named Alec who is on two committees that are measuring radioactivity all over Ukraine, figuring out which towns go in which zones. He tells us that Zone One, which is designated as being any place having more radiation than Zone Two, was supposedly evacuated in 1986. Zone Two offers three curies of strontium, twenty-five of cesium and 0.1 of plutonium per square kilometer. Zone Three has 0.15 curies of strontium, five of cesium and 0.1 of plutonium, with dosage exceeding 100 millirem per year. Zone Four has 0.02 curies of strontium, one of cesium, and 0.01 of plutonium per square kilometer, with dosage not exceeding 100 millirem per year.

Whether these numbers are correctly translated I will never know. The exceptional power of Ljudmula's right brain has left her bereft on the other side — either that or the other way around. I mean she can't say, repeat, or translate any number besides a simple integer. Decimals blow her away. We even get into powers of numbers. "Iodine-131 isotope" comes out as "131 isotopes of iodine." It's odd that she's this way. Her English is amazingly perfect otherwise, though some of the atomic terms — ion, isotope, neutron and such — are new to her. She feels guilty about this and also about the fact that she gets emotionally involved in what she's hearing. She had no idea of the extent of the horror, the depth of the complications, the forces of evil involved in the radiation.

Alec says that people are still being moved out of Zone Two, and those in Zone Three have the right to be relocated. People in Zone Four have certain economic privileges. No agricultural production is allowed in Zones Two and Three: food is being brought in to people still living there.

But of course, Alec says, things don't work out that way. People still live in all the zones, including the Prohibited Zone within thirty kilometers of the Chernobyl plant. In the absence of anything else to eat, crops are grown and consumed in all zones. Wood cook-fires send more radionuclides into the air. The Ukrainian government says that Kiev is not contaminated. Alec says that all of Kiev qualifies as Zone Four. Some parts of Kiev even qualify as Zone Three.

The trouble is, people living in Zone Four are entitled to certain benefits, including exemption from income taxation. If Kiev is in Zone Four, the government of Ukraine, already all but broke, will lose a major source of revenue.

Conclusion: Kiev is not in Zone Four. It's as clean as can be. It's just a matter of ignoring the numbers.

Alec explains how bathrooms pick up more radiation because radioactive water from the Dneiper runs through the pipes, which pick it up and hold it. He shows me a little dosimeter. For 110 rubles I could buy one in any store, he says. Then I'll always know for sure. We turn it on. It says 0.13 milliroentgens per hour. 0.10 would be normal. Outdoors, he says, it would be 0.14. This is called background radiation. It's always there.

Ljudmula and I go to a press conference back at the International Conference. No good news there. Genetic abnormalities are up 1.8 times. Cancer is way up, especially in children. But most of what I hear is the hogwash of people with political aspirations. When Elena arrives, we leave to go see a city deputy named Skripka.

Skripka's a former plant physiologist. He hasn't got any good news either. He's got all kinds of data and it looks organized, comprehensive, and accurate. I trust it because it's coming from a former plant physiologist who is wearing a suit and tie that are up to Western snuff. His data is on computer print-outs.

Skripka's theory is that there are more victims of Chernobyl than is officially known. Some 30,000 liquidators and evacuees have registered as victims, but Skripka believes there are 20,000 more who haven't bothered. But even those numbers are low, he says. Everyone in Kiev — 2.6 million people — is a victim because radiation is much higher than officially acknowledged. His data falls in line with that of Alec the chemist. Though Skripka cannot verify it, some scientists he knows went to the Soviet Union's nuclear testing ground and took radiation readings at ground zero. The readings were lower there than at any place in Kiev.

By Skripka's estimate, in the days following the accident, everyone in Kiev received from three to five rem from internal radiation. (Five rem is the maximum annual dose allowed nuclear workers in the United States.) The most dangerous internal doses are alpha and beta particles emitted from radionuclides. Alpha radiation is relatively harmless as long as the source has not penetrated your corporeal fortress. They can be stopped by a sheet of paper or human skin. Beta particles can barely penetrate skin. But if the radionuclides that emit these particles are inhaled or swallowed, the body may accept some of them as nutrients. Strontium-90, for example, behaves a lot like calcium. It has the same number of valence electrons in its outer shell. Bones readily latch on it. The strontium makes itself at home and begins radiating the local marrow.

Similarly, iodine-131 is quite like regular iodine, so it tends to accumulate in the thyroid. (The purpose of iodine pills as a safeguard against radiation is to fill the thyroid so the radioactive iodine just passes out of the body as an unneeded nutrient.) Cesium is versatile enough to find many homes in the human organism. Once lodged in the body, these radionuclides keep emitting radiation, attacking the thyroid, the marrow, the blood that happens to pass by. It might give you thyroid cancer, leukemia, or any of various blood diseases. A barely visible speck of plutonium in your lung is enough to give you lung cancer.

Kiev was also bombarded by gamma radiation. It zaps right through the human body, doing some damage on the way but not lingering to continue the attack. Today, scientists cannot determine how many rem of gamma waves people suffered six years ago.

Skripka gained access to some KGB files and found that the Soviet government knew a lot more than it let on. Children under the age of one had an average dose of .5 rem to their thyroids —half the current annual dose allowed to hit non-nuclear workers in the United States. Pregnant women and their unborn were known to have equal radiation levels in their blood. Breast milk was contaminated.

Radiation wasn't the only poison blown out of the Chernobyl reactor. The atomic chaos of the meltdown created virtually every possible element and isotope, not to mention bizarre molecules. Some of the isotopes lived for mere nanoseconds; some will be around for millennia. The lead that helicopters dropped into the flaming crater evaporated and blew across the countryside. It finds its way into people via grass that cows eat. It is impossible to say how much of this lead is from gasoline, how much from Chernobyl, but whatever the source, it's all over the place.

A special well dug in the Prohibited Zone brings water up from thirty meters. Lately it's been showing radiation levels of ten picocuries per liter. If it were a by-product in a nuclear laboratory, you'd have to take special measures to dispose of it. Dropping it into a well would not be appropriate. The Dneiper River, which runs through the center of Kiev, once had similar levels of strontium, but it has dropped to one picocurie — acceptable though still above the ideal of zero.

So it's no wonder blood donor data shows that eighty percent of donors have abnormal levels of such things as white and red blood cells and immune proteins. It explains why thirty percent of children can't receive a vaccination, because they come down with the disease the vaccination was supposed to prevent.

Skripka has data on increases in health problems. Among official adult victims, the death rate has increased 400 percent since 1987. Death by cancer is up 300 percent. Breast cancer is up twenty-six percent. General disease, up 500 percent. Problems in thyroids and other glands, up 400 percent. Respiratory disease, excluding cancer and tuberculosis, up 2,000 percent. Pneumonia, up 220 percent in adults, 260 percent in children. Allergy problems, up forty-one percent in adults, eighty percent in children. Incidence of brain cancer, up 350 percent from 1988 to 1991. Genetic aberration, ten to thirteen times higher in contaminated areas.

This information flies in the face of the conclusion of the United Nations International Atomic Energy Agency (IAEA). The IAEA made a supposedly comprehensive and unbiased assessment of the aftereffects of Chernobyl to see if people who have not been evacuated are suffering consequences of radiation. (They did not examine liquidators or evacuees.) The IAEA conclusion: no problem. There is no illness, no cancer, no increase in birth defects. Even people living in the most contaminated areas, the areas marked blood red on contamination maps, are suffering no more than they did under Stalin, Kruschev and Brezhnev.

Skripka says what the UN has done is certify the safety of nuclear war. If their report is true, he says, if fifty tons of nuclear fuel can be thrown into the atmosphere without harming anyone, why don't we dispose of nuclear waste the same way — just stack it on a pile of dynamite and blow it up?

Skripka announced a press conference where he planned to disseminate his information, which was probably the most complete and accurate available anywhere. Unfortunately, no one showed up. He suspects they were afraid to. They depend on the government for their paper and other supplies. The government of Kiev might like to have this information made public, but the government of Ukraine would not. He says he has many enemies on his committee. They don't like this information. They're bureaucrats, and it's information like this that makes heads roll.

He shows us a colorful map of topographic-like lines that show where the cesium, plutonium and strontium lie in Kiev. Ljudmula and Elena check out the milliroentgens where they live. They're in Zone Four, which is somewhat of a relief to them. It could be worse.

* *

Mr. Skripka remembers the days following the explosion at Chernobyl. He first read about it in a minor item in the newspaper on Monday, April 28, two days after the explosion. There was no mention of radiation or evacuation. But rumors were bouncing around Kiev. On April 30, Skripka checked instruments in his laboratory. They showed very high readings. Being a scientist, he knew what they meant. He called home and ordered his wife to bring their daughter inside, to close the doors, to change clothes and leave the dirty clothes outside, to take a shower, to tell the neighbors.

The neighbors didn't believe him. It was such a beautiful spring day. The sun was out, the breeze as nice as could be. The TV showed the May Day games and parades. Girls in marching bands were prancing around in short skirts, unaware of the deadly particles showering down on them. An international bicycle race charged off through the radioactive countryside, bikers huffing and puffing through the radioactive dust.

Skripka found an old Geiger counter in his closet, a relic of his days in civil defense. Somehow he got it working. It confirmed what he feared. Everything was hot, including his daughter's beautiful hair. He and his wife cut it all off with a pair of scissors. He kept calling people to warn them, but they told him he was crazy. As he watched the world continue normally outside, he began to wonder if it was possible for one person to be sane and all the rest not.

As rumors spread in the early days of May, TV announcers actively denied problems. Peasants were interviewed. They said they felt as healthy as ever. Scientists took readings near the power plant and found radiation levels *below* normal.

The news arrived in other countries first, though it tended to change fast. In some countries, it took a week to evolve from "There is no danger" to "Do not let children play outdoors." By the time the Soviet people had an idea about what was happening, there were already riots in Rome and Athens. The news was especially slow coming into France, where the government was maintaining that the radiation had pretty much stopped at the border with Germany. Like green grass, it was much higher on the *other* side. The French Director of the Radiation Protection Agency went on record saying the highest elevations were five to ten times normal, but by mid-May he was admitting he had known they were 400 times normal.

On May 6, the panic began in Kiev. Desperate to leave the city, people mobbed the train stations — the worst place to be. Radioactive trains arrived pulling radioactive dust behind them. People with radioactivity in their hair and clothes were radiating each other. They would have been better off at home because the radiation levels were already decreasing.

On May 8, Hans Blix, Director General of the IAEA, held a press conference. He said that radiation levels at the perimeter of the thirty-kilometer zone had been 10-15 millirem per hour at the time of the accident but had declined to 0.15 millirem, a safe level. He did not mention that the level in Kiev, sixty miles away, was still at 0.4 millirem.

On May 12, Kiev City Hall announced that radiation had returned to normal. The definition of "normal," however, was being adjusted for the emergency situation. Instead of 0.5 rem per year, it would thenceforth be ten rem per year. The new standard would let everyone relax because it would now take five years instead of two months to take on a dangerous dose.

CHAPTER FIVE

Vultures

Elena and I spend the day doing our duty, delivering a letter and a hundred dollars to the cousin of a Ukrainian-American and visiting somebody else's in-laws. At each stop I gather more horror stories about how it was in May, 1986.

The cousin, Mikhail, recently retired, was a foreman at a factory when Chernobyl blew up. He heard about it three days afterward. Since he was in charge of the factory's civil defense unit, he was ordered to break out the nuclear war equipment and monitor radiation levels around the factory. They began on April 30 right after lunch. As he recalls, it was .5 milliroentgens outdoors, a meter from the ground. Indoors it was .2 milliroentgens. Of course he had no idea what these numbers meant and only the vaguest idea what radiation could do to a person. He knew something serious was happening because his sister, a kindergarten teacher, told him the children of high-ranking Party leaders were suddenly taking vacations in

Moscow and Crimea.

During early May, Mikhail said, news leaked into Kiev and rumors flew. People heard about the helicopters dropping sand into the white-hot crater, efforts to pump liquid nitrogen into the pool below the reactor to cool it; to tunnel under the reactor to shore it up before it collapsed and burned into the earth; to erect some kind of protective building, a sarcophagus, to retain what was left of the fuel. There was a threat of the remaining fuel reaching critical mass and blowing up. It might also have melted down, burned its way into the ground and hit ground water to explode in a massive cloud of radioactive steam.

The panic began in the second week of May. People at the factory were asking for time off, but they were refused. They were told that all those failing to report for work would lose their jobs and apartments. Still, people thronged to the train stations and airport. A black market for tickets sprang up overnight. People were leaving town in any direction they could go — even north, toward Chernobyl.

All Geiger counters and other radiation-reading equipment were confiscated by the military for "adjustment and improvement." Mikhail still had one, however, because he had been assigned the job of touring Ukraine to measure radiation levels. He and a partner always took six bottles of red wine to flush the radiation from their bodies.

The countryside was a pitiful sight. Despite beautiful spring days of sunshine and blue skies, few people were outdoors. Radiation in the air tended to average about

.4 milliroentgens per hour. Plants were higher. Pine trees along the road measured .5. Once they saw a peasant boy tending to a dusty calf in a field. The calf and boy were both emitting .4 roentgen. Fully aware how ridiculous they sounded, Mikhail and his friend told the boy to go wash his calf and while he was at it, to take a bath himself. When Mikhail got home, his pants were giving off .25 milliroentgen.

(The .4 milliroentgen per hour radiating off the pavement was almost half the *annual* dose allowed people in the United States. *Dose,* however, is significantly different from roentgens. The *roentgen* measures radiation exposure. Much of the radiation expressed in roentgens would not penetrate clothes or skin. The *rad* measures the actual dose rate of radiation that affects (i.e. deposits energy in) living tissue. The *rem* measures dose rate, too, but more accurately than the *rad* because it accounts for the type of radiation. The *curie* is a rate of radioactivity, that is, the rate at which a material is disintegrating. The more curies present in a given surface area, the more radioactive material is present. Again, neither the roentgen nor the curie is directly related to the effects on the human organism).

Mikhail said that people in Kiev had little if any understanding of roentgens, rads, rems and curies. They had more personal concerns to contend with. Personal moral standards were spiraling downward. Everybody had a sore throat. They also had headaches because everybody was drinking vodka and red wine to make himself impermeable to radiation. Men were afraid of becoming impotent.

Nobody knew for certain that he would be alive in three months. Women were allowed abortions four months after conception, an extension of the legal limit of two months. In July, all children were shipped out of town to Young Pioneer camps in Crimea. With the kids gone, death imminent and everybody drinking red wine and vodka, Ilya says, personal moral standards went straight downhill.

Gallows humor proliferated. People joked about bearing children who could speak two languages at once — one from each head. They joked about using tweezers to pluck radionuclides from their carpets, about the dangers of washing cats and trying to get a dog to lie still while you ironed him. It was safe to eat fish from the Dneiper as long as you buried the frying pan afterwards. To make chicken Kiev, all you had to do was grasp the chicken firmly by the neck and wave it out the window for a few seconds.

Food was indeed radioactive. The government claimed to be monitoring radiation levels, but it was lying about this, just as it had lied about everything else for the past seventy-five years, and especially for the last few weeks. People came to Mikhail with their fruit, bread and milk. Risking imprisonment, he measured grain and strawberries giving off .2 milliroentgens. Milk and ground meat were less radioactive because farmers could mix clean and contaminated products to lower the radiation per kilo. You had to be careful what you bought in the market. The mere availability of food aroused suspicion. When Mikhail told people their new-bought produce was contaminated, some threw it away; some just sold it to someone else.

Elena remembers those days. She was in high school and working part time in a blood laboratory. The requests for abortions skyrocketed. In the autumn, when the radioactive leaves fell from the trees, they had to be swept up and taken away for burial. Schoolchildren were called out to do the job. They gathered them in their arms, held them up against their chests and faces as they carried them to trucks. It was fun, better than studying in the classroom, but Elena refused, telling people she wasn't a beast of burden. School officials threatened to throw her out of school, but nothing came of it.

Mikhail has a calendar on his wall. It features Lenin holding forth his fist — a typical encouraging pose. But in this one, his thumb is tucked between the knuckles of his two longest fingers. When Elena sees it, her eyes bulge, she covers her mouth with her hand, and she looks away. Apparently Lenin's gesture is not only vulgar — Elena refuses to tell me what it means, though I can easily guess — but also, until a few weeks ago, legal evidence of treason.

All this time our driver, Volodya II, is out in the car, undoubtedly shivering in the shower of damp snow. Despite Mikhail's insistence that we have more tea and cake, we set off to visit the parents of the Ukrainian in New Haven who had put me in touch with Ljudmula.

The mother and father — Ida and Ilya — are lovely, smiling, happy people. They greet us as if we are their daughter and son-in-law transmigrated from the other side of the world. Having been advised of my visit, Ida has already neatened up a bedroom for me. It nearly breaks her heart to hear I've made other plans. Unable to contain her joy, she talks and talks though she seems to know I can't understand her. She just can't stop herself. Her hands flutter in the air, land on her heart, then flutter off again. Elena translates the essential meaning of what she says.

Ukrainians are very hospitable people, but this couple goes all-out to feed their surrogate offspring. They can't find room on the table for all the breads, pickles, cheeses, beets, sweets — even butter and a bottle of Moravian port wine. Elena is aghast to see it all. "They are very rich," she says.

I take notes as I stuff my face. Ida is all but crying as she tells me how life has changed since Chernobyl. The Poleise area, a vast, virgin forest west of the plant, a place that sounds to me like Vermont or Idaho, is now too contaminated to inhabit. Ukrainian people, once true lovers of nature and the outdoors, who savored wild berries and mushrooms, are now afraid to walk in forests or even city parks. To lie in a field of grass is to risk leukemia. Swimming in the Dneiper is prohibited. Beaches along the river are closed because the sand is radioactive, not to mention that sunbathing only adds cosmic radiation to one's dose. The Soviets, having raped Ukraine for three quarters of a century, have finally robbed them of the simple pleasures of fresh air and sunshine.

Death seems rampant, Ida says. It seems more people are dying before they reach the age of thirty. Children are sick because they can't risk drinking milk. Everyone is weak and sickly. Everyone is sad. And of course she herself is especially sad; she misses her daughter, son-in-law and grandson who moved to the United States because Ukraine is no place to raise a child.

Ida has friends from Pripyat. They live in a special apartment complex built for the evacuees. Ida promises to call her friend and set up an appointment for an interview.

She and her husband sing a bit of a romantic song by a Ukrainian balladeer. Elena translates the beautiful part. It's a lot of "Darling, dear, sweetness, honey," etc. I pretend to appreciate the beauty of it, but Elena tells them that in English it is very old hat.

On the way back to town through the dark snow, sitting together in the back seat of Volodya II's car, Elena and I quickly find ourselves distant soulmates. We are of like mind in our scorn of materialism and our desire to get at the truth of things. She's about the only person who has ever understood my feeling that advertising in the Western world is very much an organized form of propaganda. Like Soviet propaganda in intent, it is dedicated to persuading people to accept the economic system. Soviet propaganda, as I have always understood it, urged people to work more. Western propaganda urges people to consume more. Quite naturally, the Western propaganda has an easier objective.

We talk a lot about books. I have to keep admitting that I don't know much about Russian literature. We have a hard time finding some titles I've read. The same is embarrassingly true of books in English, albeit mostly because she's a fan of science fiction. She insists I *must* have read a certain book, and she recounts the plot. I haven't read it, but it reminds me of a Russian film I saw in Brazil about twenty years ago, Satellite Number Something-or-Other. And yes, that's it! She saw it too. For twenty years I've been looking for someone who knew the name of that film. Elena remembers it except for the number.

What's most amazing is that Elena's been studying English (under Ljudmula) for only a year and a half, and she speaks it so well despite learning most of it from books alone. She must have a photographic memory for vocabulary, though the occasional word isn't quite right — the "hooligans" who wrecked a phone booth we passed, the "nymphomaniac" who chased her down the street one night.

She doesn't deny that she is intelligent, but she says her experience is limited. She has spent her whole life in Kiev. Here one is tied to one's apartment. To travel to another city, one needs a visa and a reason. But she has encountered foreign exoticism and some free English lessons in the form of Mormon and Moonie missionaries. They have descended upon the town like vultures, she says, taking advantage of Ukraine's problems to tempt people with hope. The Moonies, with their ridiculous propaganda, are the worst. Their argument, as she explains it, is that the Bible predicts a savior will arrive from the East. That couldn't be most of Asia, of course, because no savior would choose to arrive in a Communist country. It couldn't be Japan or Hong Kong because they are too materialistic.

That leaves South Korea, and who could it *possibly* be but the Reverend Moon?

"That fat Korean," she says so hard she almost spits. "We are from the Soviet Union. We are experts in recognizing lies and propaganda! Do they think they are putting wool over our eyes? We go to their meetings so we can learn some English and get some food. They tell us that Mr Moon is God and we say, yes, of course he is, it is only logical. Then they choose some of us to go to the United States for their full propaganda program, and of course we go because it is the only way to travel out of the country."

Elena tells me how it was in the days before perestroika. She remembered thinking, at the age of eight or nine, "Thank God I was born in the USSR, land of Grandpa Lenin. If I'd been born in the US — what a horror!" But by the age of twelve she and other kids were beginning to understand the difference between words and action. It was a point of moral crisis. Which way to go? Swallow it and follow the herd down the easy path, or remain ostracized? Elena took the latter route. I can believe it. I'm sure she has few friends and perhaps never a lover. She says she has decided to forego sexual passion, to concentrate on the life of the intellect. I believe her and I admire her for it, although I really don't feel the same way. She tells me she detests passion; she considers it an attribute of cavemen.

Elena hates the pornography that is expanding to fill the space opened by perestroika — more evidence of the vulgarity of her people. She loathes the prostitutes who sell their bodies for hard currency or marry bureaucrats so they can live well without working. At a discussion back at Ida's place, somebody tossed out the question of whether it would be better to be a scientist in Ukraine or a waiter in a rich country. Elena and I were the only ones who thought it would be better to be a hungry scientist than a well-fed waiter.

But that doesn't mean I think passion can be swapped for the pleasures of the intellect. We don't argue it, though, maybe because we're pressed so tightly together in the back seat of the cold car, leaning into the cushion of each other's coats, maybe a little more than necessary, or maybe it's only me.

CHAPTER SIX

Something Odd Going On

Lo and behold, there's a Greenpeace office in this town. I find it only by chance when Volodya (the original, back from his visit to his wife) takes me to his doctor's office. The doctor, Volodya later explains, is a bureaucrat and thus would not part with any information. But, speaking no English, he brings in a doctor in a gingham dress who speaks French. We converse. I do an excellent job of pretending to understand her. Then a young doctor comes in who speaks English. He confirms what I thought I'd heard in French, that that particular clinic has no information. Only occasionally individuals come in with problems possibly related to radiation. But he mentions Greenpeace and gives me the address.

Greenpeace occupies a couple of rooms at the Hotel Kiev, a clean, well-lit place with comfortable furniture in the lobby. The rooms are equipped with a couple of computers, a copier — the first I've seen in this country — and a television set with VCR. Of course they're surprised to see me pop in and announce the reason I'm in Ukraine. Unfortunately, they haven't got much more information than anyone else. Their *raison d'etre* in Kiev is to run a clinic for victims of Chernobyl. A young woman named Regina (the G is pronounced hard, as in *good*), who speaks beautiful, rapid-fire English, agrees to give me a tour of the clinic and the hospital where it's located, not today but tomorrow. Meanwhile, I can watch a couple of their video tapes.

One is produced by the UN's International Atomic Energy Agency (IAEA), touting its comprehensive scientific investigations in contaminated areas. It shows the radiologists setting up equipment, taking soil samples, examining leaves, poking around homes. It shows doctors feeling kids' throats, measuring people for this and that. They formed three teams: the historical team was to figure out what happened, minute by minute; the radiation team was to assess contamination of soil, water, food, houses, etc.; the medical team would measure radiation and its effects on people. The team's declared purpose was mainly to examine the validity of the official Soviet methodology and verify its findings through field sampling.

So I'm watching and taking notes when a girl named Olga turns from her computer and asks me what I think about what I'm seeing. I say it looks like they did a pretty thorough job. She says it wasn't like that.

I hit the pause button.

Olga was part of the UN team, a translator. She says the sampling was pretty scanty and that the team went only into areas suggested by the Soviets. Their sampling verified the Soviet data, but she thinks the Soviets were careful about which data they revealed. Most of the investigation was just a matter of looking over the masses of information and deciding that it somehow looked right. Some of the doctors on the teams refused to finish the job because they thought it wasn't being done scientifically or independently.

Not only did the IAEA find *no aftereffects from the accident*, but it also said that criteria for evacuation — the radiation levels that establish Zones One to Four — are too restrictive. Moving people out of even the most highly contaminated areas would do them little good; wherever they went to live they would pick up almost as much radiation from normal background sources like the sun and radon gas.

Olga has seen evidence to the contrary. Her mother is a doctor in a trauma center. From 1990 through 1992 more and more liquidators have been coming in with mysterious complaints. They have terrible headaches, sudden changes in blood pressure, inexplicable fevers. Traditional treatments not only don't work but even produce contrary reactions: raising blood pressure instead of lowering it, or worsening the headaches. There is no statistical data base from which to draw conclusions, but any doctor can confirm that something different, something *odd*, is happening.

The trouble is, these vague but very real and even

deadly complaints could just as easily result from good old-fashioned chemical pollution — the lead in the milk, the pesticides on the crops, the fertilizers in the water, the hydrocarbons in the air. Just about everybody smokes, which means any given room holds a haze the color of tar vapor; even kids must suck in a risky dose. Worsening the problem, virtually no one is eating a balanced diet. Milk is all but impossible to find and too suspect to drink even if you could find it. So everybody lacks calcium, a deficiency which happens to affect the immune system. When a kid contracts measles from a measles vaccine, is it because of radiation or the lack of calcium caused by a *fear* of radiation in milk? I guess the question's moot to the kid who has measles for three months.

I don't know where people find their food. I haven't seen any for sale. Up the street from the Environmental Protection Society, there's a state store that has bread every couple of days. It's dirt cheap and not bad, and for a few hours in the morning bread is in good supply. You can buy full, half or quarter loaves right off a tray, unsliced, unwrapped. I carry it home in my coat pocket. Along the way I pass a place where street dogs sleep, curled up on steaming manhole covers. Sometimes two or three dogs share one of the round steel beds. With their fur blended together and their noses tucked under their tails, they look like Russian hats. I always toss them a piece of bread, but they regard it with hesitant caution. They don't get up, probably for fear of losing their place, but they do keep their sleepy eyes aimed at it.

The store offers little else. There are gallon jars of pickled green tomatoes, gallon jars of brownish juice which I've been warned not to drink, little jars of strawberry jam, and that's about it. The rest of the store is cold concrete and empty shelves. The lady who takes the money really couldn't care less whether I pay cash or choke to death. Of course I can't quite understand the prices — nothing's labeled — so I tend to hand over a whopping ten rubles and wait for my change. But ten rubles is *way* too much for something that costs mere kopecks. Looking terribly, terribly miffed, the lady bitches at me, slams her register shut, crosses her arms over her chest, looks into the distance and waits for me to get out of her face, wearing an expression like Lenin looking over a tractor factory. Once I just put a nice pile of kopecks on the counter and walked out with my quarter chunk of bread. Apparently there weren't enough kopecks. The lady chased me up the street, squawking words I wouldn't repeat in public even if I could.

Bread is about the only food I can find. I live off the canned stuff in my suitcase — tuna, peanut butter, granola bars, coffee — and whatever the generous Ukrainians offer me when I interview them in their offices and homes. Sometimes when I'm out all day I don't eat a thing for ten or twelve hours, which is probably a pretty standard situation in this country. But I get used to having the raw gnaw of hunger in my stomach. For some odd reason, I almost like it.

I brew up tea or coffee with an immersion heater stuck in a glass. Andrew lent me an aluminum kettle with an immersion heater in the bottom, but I use it to hold water. Making coffee is a messy process, involving a trip to the indoor outhouse downstairs, where I wash a couple of glasses with a bar of bath soap. Since the coffee isn't instant, it involves grounds and a dirty filter. Of course I have to offer some to everyone who visits my room — Kiev has run out of coffee — so I repeat this process several times a day.

One day there is meat in the store. A dense troupe of babushkas huddle in a broad, deep line at the counter. I walk in with the angelic, blue-eyed Vyka, Andrew's partner in film projects. She, like everyone else in Ukraine, needs cigarettes. Since we aren't there on meat business, we elbow our way through the babushkas . . . and behold! Behind the counter, right up there with a line of pickled green tomatoes, is a low pyramid of "Milwaukee's Best" beer, sixteen rubles the can. The butcher tells me it's German. I manage to convince myself it's true — I would rather have good German than cheap American beer — and buy five cans, stashing them in five coat pockets. Vyka pays for it but then won't take money from me. She says it's a Christmas present. I feel very guilty walking down the street all lumpy with beers paid for by somebody who surely must consider fifteen cents real money.

So how do you figure that? How does American beer sell here for fifteen cents a can? Haven't they grasped the concept of profit? They ship the stuff over here by God-knows-what tortured route and sell it for surely less than the US wholesale price, maybe even lower than the production cost. They haven't grasped the first lesson in Capitalism 101: Buy low, sell high. No wonder the country's going broke.

CHAPTER SEVEN

Chernobyl AIDS

A couple of mornings before Christmas, in the pitch dark of 7:30, Volodya honks his horn under my window. I go downstairs and let him in. He asks for his entire week's pay in advance, and then some. He needs a hundred dollars, US cash, to fix his shot bearings and a host of other ills ranging from finicky brakes to no benzene. If I ever saw a set-up, this is it. If I give him a hundred dollars, I'll never see him again. But on the other hand, how can I say no? The guy saved my butt. He gave me his home. He didn't kill me. Do I want to drive around in a car with no benzene, no bearings, no brakes? No way. So I fork over more cash than a successful prostitute makes in a month, figuring that if he runs off, I'll have Ljudmula hunt him down and attack him. She's been warning me that Volodya and Andrew are untrustworthy types. She can tell by their rotten Russian and the sleazy look in their eyes. But I'm not one to say no. He's supposed to be back at two o'clock to take me to a children's hospital where Regina, of Greenpeace, has set up an appointment for me to meet with some doctors and take some pictures.

Two o'clock arrives but Volodya doesn't. Plotting his murder, I huff off to find a taxi — no mean feat on a sloppy day in a town that's low on fuel. By "taxi" I mean any given motorist who wants to make a little dough. What you have to do is hold your hand out until somebody stops. This takes quite a while because even on a main avenue in downtown Kiev, only a couple of cars pass in a given minute or two.

But one finally stops. The driver's a young guy. We don't talk about price or anything. I show him the address of the clinic and off we go. Then he says in English, "Please fasten your seat belt." By gosh, it turns out this isn't just your average motorist hoping to make a little dough. He's the World Champion Trampoline Jumper. He won the title in Birmingham, Alabama, and therefore speaks a little English. I give him thirty rubles for the ride, which is cheap considering the privilege.

I get some good information at the children's hospital. A calm, exhausted doctor named Olga tells me that Ukraine, as a whole, is in terrible health. She does not claim any significant increases in cancer, leukemia or birth defects. The problem is in the general state of health. Everybody, especially children, has a *generally* poor state of health. *Everything* in their bodies is breaking down and there is no single explanation for it. This problem has certainly worsened since Chernobyl, and especially over the last couple of years. But Chernobyl is not the only culprit. The pollution and malnutrition combine to weaken and attack the body. Just as important, perhaps, is the stress of knowing that all this is happening to everyone, plus the stress of unemployment or chronic underemployment, plus not knowing if there will be any food in town, let alone in the cupboard, by the end of the winter, let alone by the next harvest, plus the inevitable family problems produced by hunger, illness, poverty, alcoholism and living in close quarters, plus they really don't even know what kind of a country they live in or how to replace the ruins of the centrally planned economy, plus several of the other former Soviet states are breaking out in civil war, which could happen in Ukraine, which is artificially attached to Crimea, which doesn't like being Ukrainian. It's all enough to make anybody sick.

The combination produces what Olga calls the synergy effect. The sum of the ills is greater than the combined normal prognoses. In fact, the sum of the ills is an illness nobody's ever seen before. It varies from individual to individual. It's as if each person had a new and unique disease. You can't tell which is deadly, which will cure itself. Measles isn't just measles anymore. Treatments don't yield normal results. Textbooks no longer apply. The medical progress of the twentieth century doesn't count for much. They're starting from scratch.

What they're dealing with is a syndrome resulting from a massive attack on the body's immune system. The body, starved for nutrients, is glad to latch onto whatever elements come onboard, even if it's cesium, strontium or plutonium. The isotopes make themselves right at home, radiating the immediately surrounding areas and all the blood that passes by.

This isn't an illness; it's a syndrome. It's an immune deficiency, and it's acquired from the environment. Put it all together and it spells AIDS. HIV doesn't necessarily enter into it (though often enough it does), so people refer to it as Chernobyl AIDS. Unlike HIV, it isn't contagious, but like HIV, it has no cure.

The real problem, properly defined, isn't the suppressed immune system but the unpredictable responses to treatments. The diseases are unique and the treatments are unknown.

Olga thinks the solution lies not just in building another hospital or institute, but rather in recognizing the problem and developing a new medical specialty. Of course no such action is being considered in Ukraine or anywhere else in the world.

Another doctor, dark and serious, tells me that this new medical information is not completely new to the world. A considerable body of data exists in the Moscow Scientific Institute, but it is classified, so no one has access to it. The data comes from victims of a nuclear accident at Chelyabinsk and also people subjected to nearby atomic tests. This doctor saw secret papers referring to this data and had already deduced its existence when Moscow sent investigators after the Chernobyl catastrophe. From the questions they asked, it was clear they had encountered similar problems in the past. They already knew things it would take Ukrainian doctors several years to figure out through trial and error with thousands of human guinea pigs.

Doctors call this "catastrophe medicine" — the necessary use of patients for experimentation, the direct application of theory to practice.

What Olga and her colleagues have done at this particular hospital is develop a computer program that analyzes *all* of a patient's functions to produce a single holistic treatment for a given individual's multifarious ailments. It works well, or would if they had the medicines the treatments require. But this is a hospital without medicines. They don't even have a ribbon for the computer printer. They show me a print-out. It's too faint to read in a room that's down to its last light bulb.

Olga lets me walk around to take pictures of half-dead kids who are conceivably radiation victims. One is a very small girl with acute immunodeficiency which has led to kidney failure. A fat little baby suffers multiple genetic abnormalities, including Down's Syndrome and lack of anus. Another is a boy just withering away for reasons unknown. Another boy was doing fine until his father got drunk and smashed up his car with his son in the passenger seat. Now the child is in a coma. A psychic is with him, waving her hands over him as if caressing an invisible essence outside his body. A doctor says the boy showed a little response, a movement of a leg. The doctor doesn't look optimistic. He says the psychic is a last resort.

I take a subway back to my room at the Environmental Society. The subway isn't hard to master. The system is rather simple, just a couple of lines crossing each other and a couple dozen color-coded stops. The map, unmarred by graffiti, is in the same place in every car. But what I find difficult is extracting three five-kopeck coins from a change machine. Clever me, BA, MA, MA, MFA, I finally figure out how you put a 10- or 20-kopeck coin into the right machine (different machine for each coin) and make the five-kopecks shower down, just as it says, no doubt, on the front of the machine.

Then you can plunk the right change into the turnstile, which, unlike New York's, has nothing to turn. But woe unto him who tries to walk through without paying! You trip a beam of light and a steel gate *leaps* into your path, right at groin level, like something from an Indiana Jones movie. It's not a mistake you make twice.

The escalator is child's play for a deft young guy like me. It rushes forth at twice the speed of its slowpoke cousins in America. You have to step onto it in a forthright manner. Then you're decidedly on your way. There's no going back. It goes way, way down to a depth that is supposed to withstand atomic explosions. You can't see the bottom from the top, and despite the speed, it seems to take a long time to get there. You have time to get bored. One day I got so bored I reached out and flicked at the little black knobs that keep kids from sliding down the stainless steel areas between the escalators. When I hit the knob, it twisted to one side and the whole machine stopped.

When I twisted it back, the machine didn't start. It was the upbound escalator and it was during lunch-time rush hour. Several thousand people had to walk up a hundred yards of stairs, all at the pace of the slowest babushka in the line. Me, I was four steps from the top and made a quick getaway. I daresay it was the fastest I ever left a subway.

As long as you know what station you're at and the one you're going to, you can do it. Trouble is, the signs aren't easy to read. They're in Cyrillic, of course, and I don't know the names of any places except the Oh Baloney stop, where Elena lives, and what sounds like *"Stadio Centralee"*— Central Stadium — which is near where I live. At a glance, any given sign might be the name of a station or a warning not to spit. But with enough time and cogitation I can sound out the letters and compare them with other signs and what I hear over loudspeakers. I don't spit and I wait until I hear *"Stadio Centralee."*

Actually I'm sure it doesn't say not to spit, as it does in every subway car in New York. No one would spit in a subway station that has chandeliers, statues, marble floors and digital clocks that tell you how long since the last train left — never more than two minutes before the next will arrive.

People have warned me not to go out alone, especially after dark. But it seems as though it's always after dark here, and I'll be damned if I'm going to work only until three o'clock and then scurry home.

So when I'm on the subway or walking down the street, I do it in a forthright and confident manner. Walking, I move fast and keep checking behind me. Standing, I keep my feet well spread, my knees slightly flexed. I look from person to person to person to try to figure out who might have murder on his mind. I also look for what it is that sets me apart as a foreigner. Two things are my nice new hiking boots and my commodious coat, two items for which people will literally kill. Another is the fact that I do not carry a plastic shopping bag like everybody else. But except for the few thousand people who had to walk up that long, dead escalator, no one has shown any interest in murdering me.

CHAPTER EIGHT

Merry Christmas

Later that night, Volodya returns to fill my room with the smell of rotten egg and the gleam of his golden tooth. "*Machina* very good," he says with a kiss of his fingertips. "Oh, very, very good."

It's a relief to see him back. It hurt to think I'd been ripped off, and now I hurt to think I'd thought ill of him.

The next day is December 25 — no big deal around here. Russian Orthodox Christmas falls on January twelfth, and the bigger holiday is New Year's. So Volodya drops me off at the hospital that houses the Greenpeace clinic and a leukemia ward.

Regina, who reminds me of certain tall, dark Brazilian women I've known, takes me for a tour. Ho, ho, ho, merry Christmas: skinny little kids are dying of leukemias so strange and complex that doctors hardly know what to do and even if they did, they wouldn't be able to do it for lack of equipment and medicines. Many mothers are there with their doomed children. Most allow me to take pictures though they're embarrassed that they're not dressed for the occasion. One little girl, very pretty, motionless on her bed, has no chance of recovery. Does she know that's why I was taking her picture? I think she is too weak to ask, but her eyes follow me as I shoot her from various angles. She lies curled up in a red print dress, a kerchief over her bald head. It is too much. I start to cry.

The doctors are too professional to link hematological cancers to Chernobyl. But worse will come, they say. Leukemia tends to develop six or seven years after exposure. In the next year or two the radiation will have accumulated in enough bones to fill enough beds to point at the undeniable cause of it all. With a little luck, the situation will be so bad that governments and the nuclear industry will no longer be able to deny that something's amiss.

* *

Next morning comes a knock on my door and in, unbidden, strides a heavy-set man in a gray trench coat. He seems to know me, jabbers away, seems to be saying that he's to stay in the room next door but has no key as of yet. Can he leave his stuff in my room? Of course he can. How about his three friends? Who am I to say no? They come in and out, in and out, as they freshen up from their bus ride from the Crimea. Then they ask — all this with gestures and nods and bus noises — if they can use my kettle. Of course they can. And with that they proceed to make themselves at home, clearing off my desk and spreading out a loaf of bread, a cake, and other breakfast goodies, sitting down and gobbling it all up. I decline their invitation to join them. It's quite an odd situation. I have to leave and would like to lock my room — not that I don't trust them (and maybe just a little bit I don't) — but if I leave without locking my door, it will be unlocked all day. I try to explain, but they mistake my gestures to mean that I want them out, which isn't *exactly* the truth. But pretty soon they've washed everything and disappeared into their room. Volodya shows up a while later, and my new next-door neighbors come in and hand me a liter of vodka.

* *

I spend an afternoon as a guest speaker at Ljudmula's English class at the Institute of Foreign Language. The room is small, harking back to the days when classroom chairs and floors were wooden and the blackboard was a slab of thick slate. The students are all adults in threadbare clothes. Their questions are about economic matters. Not wanting to glorify the wealth of the West, I expound on the commercialism of Christmas and the evils of capitalism and materialism. I don't feel it necessary to go into the proven disadvantages of Communism; they have that knowledge in their thyroids and bones. I tell them *I* think they should start their new economy on an agrarian foundation and build themselves up from there.

It's futile, even dangerous, I tell them, to try to manufacture cars or computers if you can't find light bulbs for your factory or paper for your bookkeepers. The accident at Chernobyl, I could (but do not) argue, is the result of technology attempting to exceed the capacity of its sociopolitical foundation.

Someone asks me if I think there is any hope for Ukraine. I say yes, Ukrainians are intelligent and educated. They'll find a way to work themselves up. The problem, as I see it, is the bureaucracy, which seems likely to continue. I know bureaucracy from my days in Brazil. It doesn't like to relax its cancerous stranglehold. I have no idea how to eradicate a bureaucracy.

Someone else asks me whether capitalists are born or can be made. He seems serious. I explain that the reason capitalism works is that it's automatic, natural, instinctive, the law of the jungle. I tell him about the Milwaukee's Best beer I bought. I go for a big laugh, but everybody just looks insulted.

After class somebody tries to sell me information about Chernobyl. It is the same guy who handed me the limp flower for my hostess. His offer is in cahoots with Alec, the chemist. The prices are ridiculous — $20 a page or more. Granted, some of that money would pay for translation, but only a small part of it. The total is close to $1,000. I suspect they just got carried away with their dreams, wild guessing at the parameters of Western financial reality. The fact is, I get all the information I need just from their descriptions of what they have. Some of it is damning evidence of how the Soviets suppressed information, including, they say, written instructions ordering local authorities to continue to deny any health danger or aftereffects.

CHAPTER NINE

Minister and Morgue

Ukrainians aren't the only ones with weak immune systems. I weakened mine drinking a late-night overdose of vodka with the Crimeans next door. I went over to bum a cigarette and they put out a full spread of breads, meats, jams, butter, all kinds of stuff. They had a big box of butter, a good ten pounds of it. They wouldn't let me leave. These guys knew absolutely not one word in English, not even "yes" and "no." Somehow we exchanged information on some things — my purpose in Kiev, the beauty of Crimea, and so forth. Then I went back to my room and settled into a bad case of the flu.

I don't know if anyone else can do this, but I can smell the flu on people's breath. I've smelled it all over the place in Kiev and figured I would get it eventually. I do, and it hits me hard. A hangover puts a nasty edge on the queasy stomach, sore throat, headache, gloppy nose and a chest that feels like it has a knife through it. I had brought

Tylenol from the States but had given Volodya the whole bottle when he had a toothache.

Unfortunately this is the day I'm scheduled to talk with Yuri Shcherbak, Minister of the Environment of Ukraine. There is no canceling this one. Volodya shows up and takes me to a government building downtown where Ljudmula is waiting. There's a long line at the elevator, so we walk up several flights. I barely make it. At Shcherbak's office I can barely breathe.

Yuri Shcherbak, I've been told, is one of the few government officials who is honest and intelligent. He never joined the Communist party. He used to be a doctor and has, in fact, published a book called *Chernobyl: A Documentary Story*. It relates a lot of damning evidence, not as straight data but rather as personal accounts of what happened and how it affected people.

He doesn't want to tell me anything about alleged cover-ups, but he wants it known that while the official data says 50,000,000 curies were thrown out of the reactor, it was really three times that and perhaps as many as a billion. At least fifty tons of fuel went up into the air, and perhaps a lot more. The clouds of radioactivity have circled the earth several times. In eastern Europe, seven million hectares were highly contaminated. In Ukraine, 2.5 million hectares are radiating more than five curies of cesium per square kilometer. A quarter of Ukraine and a third of Belarus are contaminated. The soil cannot be used for many years to come. Maybe it will wash away to poison some other place for a while, or maybe it will hang around for a couple dozen millennia.

And the world hasn't heard the last of Chernobyl. The sarcophagus was built with many vents. Rain comes in through those vents and gradually dissolves the remaining nuclear fuel and washes it into some of the 327 offices, halls, closets and other rooms in the building. Nobody knows exactly where the fuel is or where it's building up. The only way to find out is to drill into the rooms, one by one, and send in a probe to take readings.

Last July the readings were not good. Somewhere within the building, enough radioactive material was collecting to reach critical mass. If it continued, it would blow up — not like the little steam explosion that blew the roof apart. It would be an atomic explosion like the one that blew Hiroshima apart. Such an explosion would not only throw the rest of the fuel and its radioactive surroundings into the air. It could also destroy the two neighboring reactors still operating. All told, it could throw another three or four hundred tons of nuclear fuel into the atmosphere — several times more than in the original explosion. If that wouldn't be the end of the world, it would certainly bring it within view.

But they managed to drill into the room where the radioactive materials had accumulated, and were then able to pump in something that slowed the rate of reaction. Now Shcherbak says they are *almost* totally absolutely one hundred percent positive that a similar incident cannot happen again. When he tells me that he reported this information to people in the US Department of Energy in

Washington, he shows me how they put their heads in their hands and rocked them back and forth.

The government of Ukraine has announced an international competition offering big bucks to anyone who can design a functional and effective new sarcophagus. The one they have isn't going to last as long as the radioactivity. It was built in a big hurry by people who didn't feel like taking extra pains to do the job right. The plutonium inside will be radioactive for about a quarter of a million years. The sarcophagus is good for thirty. If there's an earthquake or something, it could be less. At this rate, Ukraine can look forward to building ten thousand sarcophagi, each enclosing the ones built before.

God forbid there should be an earthquake before the year 248,008 A.D. If a wall fell in, it would heave up several more of the thirty-five tons of radioactive dust that is in the sarcophagus today.

There used to be four reactors operating at Chernobyl, with a fifth under construction. Number Four blew up. They shut down its neighbor, Number Three, for the obvious reason. Even before the fire at Four was out, liquidators were carrying away fuel rods from Number Three. The rods were stacked on platforms that looked like funeral biers. Liquidators carried them away by hand, running just as fast as their legs could carry them.

In June 1991, Ukraine decided to shut down the last two reactors by 1995, but in October of that year, Number Two caught fire, destroying the roof of the building. Then Number One caught fire when some corroded wire shorted out. Ukraine decided to shut the whole thing down ASAP, to wit, 1993 (though in 1995 it would still be operating).

The parliamentary decision, perhaps passed in a moment of hysteria, left the question of where Ukraine is to get its light and power next year.

Minister Shcherbak says they can limp along with their existing nuclear-, coal-, and gas-powered stations. Currently the country uses three times more fuel per capita than the United States, an indication of some leeway for improved conservation. The cost of electricity is low — one kopeck per kilowatt in rural areas, four in the city, something in the range of one to four hundredths of a dollar, which is cheap even by local standards. If they switch to gas, of course, it will cost more. The trouble is, Russia has cut off their supply of gas, so they're burning coal, and not just any coal but very cheap and dirty coal, creating more pollution than anybody needs.

Chernobyl itself may not be Ukraine's biggest environmental problem. The country's water supply is quickly approaching sewage quality. Thirty million people drink from the Dneiper, which not only drains the vast watershed around Chernobyl, but also drains away 1.2 cubic kilometers of sewage each year, and that doesn't include the wastes from 600 industries that dump into the river. On top of that, fertilizers long since banned in Western coun tries still wash into the river. The water for human consumption is cleaned, but twenty percent of it cannot meet standards of purity. Nobody knows how much radioactive sludge has settled in the bottom of what is known as the Kiev Sea, a vast lake just north of the city. But they'll know when the next heavy rains hit because it will all wash over the dam, float down the river that flows through the middle of Kiev and flood into the Black Sea. Minister Shcherbak has absolutely no idea how this situation can be adequately improved to the point of human tolerance.

The interview does nothing to help the chemo-biological pollution that is raging through my head, chest and stomach. I feel like a living metaphor of the Ukrainian condition.

From Shcherbak's office we go directly to the Ukrainian Academy of Sciences to talk with Academician Dr Dmytri H. Grodzinsky, who probably knows more about radiation than anyone in the country. He also has a pretty good handle on the politics involved. He delivers a long, long monologue, all of it good but not all of it directly translated by the indefatigable Ljudmula. This has been an awful lot of work for her. She's been hanging in there out of sheer anger against the people who caused the Chernobyl mess and its aftereffects. She hadn't known all the ugly stuff we're hearing, and often she gets too emotionally involved to restrain herself to mere translation. In Dr Grodzinsky's office she's doing a lot of listening but not so much talking. I assume she's filtering out what I already know or don't need to know and condensing the rest.

Grodzinsky is rather old and, sitting in the midst of glass-encased bookshelves, he looks very wise. He speaks in a careful measured way, seeming to say only what he's sure about. I have the distinct and certain feeling that this is someone who is honest and intelligent and who is going to give me information untainted by fear, anger or self-interest.

Chernobyl, he says, is very complex. It can be seen as the focal point of all that was wrong with the Soviet system. That system fostered incompetence, encouraging it to rise to the top, and in the overall Soviet business, nuclear energy was the top. The Ministry of Medium Machine Building, which oversaw the nuclear energy program, was the wealthiest ministry, and the people involved did very well for themselves.

I know from other sources that the people in charge of nuclear power plants aren't necessarily nuclear engineers. The director of the Chernobyl plant, V.P. Bryukhanov, was an expert in turbines whose previous experience was at coal-fired plants. The chief engineer, Nikolai Fomin, had been chief engineer at the Balakovo nuclear plant where, due to slipshod work by those under his command and breaches in safety rules, fourteen men were boiled alive in the compartment that surrounded the reactor. The deputy chief engineer in charge of operations at the Number Four reactor had worked in a physics laboratory and had some experience with small marine reactors but none at an atomic power plant. Apparently these three counted on their underlings to run the plant, but did not hesitate to approve an inexcusably unsafe experiment that doomed the reactor. The people at the Soviet Department of Energy who were supposed to approve the plan for the experiment didn't bother to respond to the request for approval.

The incompetence reached throughout the Soviet bureaucracy. The plant had been poorly designed and then built to substandard specifications. Since a lot of cement was stolen, the concrete was weakened with added sand. Defective metal was installed so that construction could stay on schedule. The assessment of the accident in the first hours was horrendously slow and inaccurate. The evacuation was botched. Medical consequences were hidden or distorted.

So, according to Grodzinsky, deception is the first element in this tragedy. The second is politics. Politicians, suddenly thrust into something resembling democracy, have gravitated to two extremes. Either they deny that Chernobyl is really a problem — conscious, outright lies — or they exaggerate the damages with strident hysteria. The latter have garnered much more support, though none of them has much of an idea what to do about the problem.

The third element is money. Not only does nuclear power produce a lot of money — much of the power is exported to other Eastern European countries — but the Chernobyl phenomenon itself is big business. It brings in millions of dollars in foreign aid, much of which gets divided up among the bureaucrats. As soon as the world faces the fact that a new sarcophagus needs to be built, a billion dollars will flow into the country. Hospitals hope to reap foreign aid as they offer evidence of victims and the need for equipment, medicine and supplies. All the new technologies that have to be developed will become exploitable businesses in which some people will strive to solve problems and others will strive to make money.

With all of this at stake, there are no impartial positions. Nor is there a line between lie and truth. But some people have clearly taken positions well to one side or the other. Estimates of the death toll, for example, range from twenty-nine to ten thousand. Those who claim the former are lying; those who claim the latter are either lying or hysterical. The truth is between the extremes.

Dr Grodzinsky says that among the liars are those who called in the IAEA to support their incredible claims of little or no lasting damage from the accident, and the IAEA backed them all the way. The international nuclear industry continues to ignore what it knows and goes on minimizing the negative aspects of nuclear power.

I'm pleased to hear Grodzinsky say good things about Yuri Shcherbak. He calls him honest and intelligent and really working toward the truth, though he must do so within an inherently corrupt and uncooperative system.

Meanwhile, Grodzinsky says, Anatoly Romanenko, who at the time of the accident was Minister of Health, still insists that there are no health dangers, that the supposed ill effects are just the result of stress brought on by "radiophobia." If people could just learn to relax, Romanenko says, everything would be all right. His current job is Director of the Ukraine Institute of Radiology, which is supposed to be researching the effects of radiation on people.

Grodzinsky was a liquidator, called in as one of the country's top experts on low-dose radiation. His first investigations in the vast area affected by Chernobyl's radiation were necessarily scanty and incomplete. He found a lot of cesium, plutonium, strontium and, in the first few days, iodine-131. Doses suffered by liquidators and populations were only roughly estimated and tended to be recorded on the low side. (The IAEA found Soviet estimates on the high side.) Now, says Grodzinsky, it is impossible to accurately estimate doses taken in those first few days. They have to do it with principles established in research in Hiroshima and Nagasaki, which were events very different from that at Chernobyl. The "source terms" — the radioactive produce of nuclear incidents — were different because two were atomic explosions, the third a meltdown. The effects on the population were different because the Japanese cities were populated mostly by women, children and the elderly, who are presumably more susceptible to injury. The Chernobyl area population was more balanced, but the people had probably been exposed to unknown quantities of radiation *before* the accident, from leaks and carelessness with waste. The Japanese populations were exposed to a sudden burst of radiation, while the Ukrainians and Belarusans (and Russians, Finns, Poles, Germans and others) were exposed to long-term low-dose radiation. The situation in the Japanese cities, given the extensive non-radiation injuries and the stress of war, killed off the weak, leaving the stronger to serve as the basis of statistical information on the effects of radiation.

In other words, the assumptions drawn from the Japanese experiences are less than entirely valid in other situations.

Academician Grodzinsky did extensive research on the biological effects of Chernobyl's radiation. He found many trees with cancer, a condition virtually unknown in nature. He found genetic abnormalities in farm animals, though they were quickly disposed of and denied by the government. He found pine needles ten times their normal size and oak leaf clusters with one leaf several times broader than its neighbors. This gigantism was very common in the areas within a hundred miles west, east and north of Chernobyl.

No one reaped the wheat that grew in those areas after the accident. The seed fell to the ground. In the spring, it sprouted. Within three years, eighty percent of it had mutated. The mutations took place under conditions of low radiation. Ironically, one of the mutated forms turned out to be very resilient and productive.

Dr Grodzinsky reports public health problems along the same lines that I've heard elsewhere. Thyroid cancer is up from one case in 1986 to twenty-two by April in 1991. In 1985 there were nine cases of cataracts in a village near the plant. In 1988 there were 241. Since 1988, the number of children with blood disease has doubled. In 1990, eighty-two percent of children showed changes in their lymph nodes. Only thirty-five percent of schoolchildren had normal health. About twenty-five percent of children evacuated from Pripyat have weakened heart muscles. Pregnant women are having more complications and they happen earlier in their pregnancies, though this may be because women are having babies at older ages, eating less nourishing food and suffering from the kind of family stress brought on by unemployment and unstable finances. And they don't know what kind of a country they live in or will live in, and nobody can walk in the woods or parks or lie on the beach or take a vacation.

Virtually all doctors, says Grodzinsky, agree that all forms of disease are more common. Children catch more colds and respiratory diseases, and the illnesses last longer. Cuts and burns don't heal quickly. It's hard to operate on people because the wounds don't heal.

You don't need the radiation to yield this effect. Fear of radiation is enough to do the trick. People afraid of strontium avoid milk, which means they don't get enough calcium, which is important to the healing process. On top of that, if the body is starving for calcium, it's all the more likely to accept strontium as a nutrient when it comes along.

Grodzinsky remembers when they changed the acceptable level of radiation in milk curd to 10^{-6} curies, which exceeded the level that qualified nuclear laboratory materials as toxic waste. Laboratory waste would have to be disposed of by using special filtering procedures, and burying the residue. But it was quite legal to dispose of contaminated milk by filtering it through children.

According to Grodzinsky, one reason it's hard to establish changes in general public health is that before 1986, the government did not want health statistics known, especially those indicating epidemics or cancer. A variety of deaths were attributed to acute heart deficiency, sore throat or bronchitis. This is what went on right after the accident, and still goes on even today, though perhaps the opposite is occurring now as doctors try to blame illnesses on radioactivity. The Soviet-inspired inaccuracies would explain why illness rates seem higher now. Rates also go up because today everyone is required to see a doctor once a year, which necessarily increases the rate of detection. All of this adds up to a statistical mess which has little value and needs a lot of qualifying.

Grodzinsky knows scientists who are searching for the truth about human illness by investigating illness in white rats. They found that cancer tended to develop at nine months in rats which were fed slightly contaminated food to produce an internal dose rate of 0.06 sievert. Rats fed clean food *might* develop cancer, for other reasons or from the external environment, but only at sixteen months.

The trouble is, these rats receive carefully balanced diets. Ukrainians do not. The statistics coming out of the laboratories, therefore, are skewed toward optimism.

Furthermore, experiments on rats produce inaccurate (and again, optimistic) results because rats don't live long enough. Since people live longer, they have longer to suffer the effects of low-dose radiation. However, it seems that equal doses accumulated over short and long periods do not yield the same results. You may be better off taking, say, ten rem all at once rather than over the course of several years. One theory is that the short, hard burst of radiation will tend to kill more cells, effectively preventing them from mutating. Under slower but equal dosage, the cells survive and mutate into cancerous growth. If rats lived as long as humans, it would be easier to prove this. The effects of low-dose radiation cannot be determined by simply doubling the dose and halving the time of exposure. A simplistic formula like that was used to draw invalid conclusions from the effects of radiation in Hiroshima and Nagasaki.

God bless Dr Grodzinsky, he gets me some food. I suspect I look like I need it. All I've had to eat all day were a couple of granola bars before dawn. By three in the afternoon, I look like I'm dying of radiation poisoning. To my surprise, a nice lady shows up with a tray of hot food: beets, curds-n'-sour cream, fried potatoes, cutlets, bread, and something else from the vegetable family. But it's all for me, nothing for Ljudmula or Grodzinsky. I know Ljudmula hasn't eaten all day, but she declines to accept anything. So I tuck in, albeit with self-conscious guilt. I scarf it down like a Biafran, eating beets for the first time in my life, mopping up the juice as if it were the nectar of immortality.

Nothing like radioactive beets to revitalize a guy, at least enough to make it home. It's almost dark when Ljudmula and I make our exit and go search for Volodya, who's been parked somewhere for the past several hours. I assume he's been hungry all this time, too. There are no corner bars or luncheonettes in this city where you can grab a bite. If you want a snack, you probably have to look for it on the black market. Ljudmula and I cross a treacherously icy empty lot, walk up and down a couple of streets and finally find our man parked up on a sidewalk, elbow out the window, smoking a cigarette and listening to his tape deck.

As soon as I get home I hit the bed, resolving to stay there until I die or feel better, sinking into the mattress for all of the three or four minutes it takes for two truck drivers to arrive and ask if they could sleep in the other two beds in my room for the night. Who am I to say no? They move in quietly and apologetically and set up a meal on my desk. They all but beg me to join them, at least in a dose of schnapps. I decline. I just want to sleep. I retreat to an igloo beneath my covers, but can still hear them chewing bread and slurping schnapps truck-driver style. I can smell their canned fish. Then they sleep quietly and the next morning ship out before dawn.

Still sick, I try to stay in bed, but sweet Vyka, Andrew's filmmaking partner, arrives to tell me she's lined up an interview with a certain journalist on the day after tomorrow. Then she invites me to go sightseeing on Saturday at ten o'clock.

Volodya shows up with Volodya II. I let it leak that I'm sick. Volodya tells Alla, the grandmotherly woman a couple of offices down from my room. Now she's all in a tizzy. She brings me some jam-like stuff made from bitter berries guaranteed to cure bronchitis. Also some strawberry juice in a jar. She phones a couple million people in search of aspirin, though I keep telling her I'm not worth that rare and precious commodity. What I really need is lots of liquid, but there's none to be had in Kiev except for good old radioactive tap water from the most polluted river in the world. I hate radioactive polluted tap water. Boiled radioactive polluted tap water doesn't seem half as bad, though, so I subsist on tea.

So, except to deal with well-wishing visitors, and a quick laundry job in the sink of the indoor outhouse downstairs, and typing up notes while I can still read them, I keep to my bed. I also call uptight Elena to set in motion a complicated chain of information and transportation that will result in some interviews of evacuees from the town near Chernobyl. Because of my ignorance of the language, setting things up involves coordinating a translator, a car, and the interviewee. I have to make sure Elena gets all the information straight so she can relay it to the right people. It's hard on a phone system that's not much more sophisticated than tin cans on a string. My ear begins to hurt from pressing the phone against it so hard to hear her faint and distant words. We talk for a long time while a cool draft leaks in through the window behind me and trickles down my neck.

Elena's very concerned about me, a child in a jungle, she says, "like Old Yeller." She has to repeat that name several times before I get it. She pronounces it well enough, but who's expecting such a name to come in over a phone line in Kiev? Old Yeller hasn't crossed my mind once in the past three decades. Anyway, now I'm Old Yeller with bronchitis. She wants to visit me. Shy, uptight me, I turn down the offer on the grounds that she'll get sick, too.

For lunch I open a can of chicken chunks packed in water, and a peanut butter and jelly sandwich on yesterday's bread. Alla stops in several times to check my fever. I think she tells me she's searching hard for aspirin. It wouldn't surprise me if she somehow drew a bowl of chicken soup from her breast the way good grandmothers can do.

Since the moment I arrived in St. Petersburg I've been trying to formulate an answer to the inevitable question people will ask me back home: *How is it over there?*

Well, I don't know enough to answer that. Sure, the store shelves are nearly bare; there are long lines for meat. But I have the feeling people have enough to eat so far, though certainly not a balanced or even interesting diet. Everybody looks desperate, but nobody looks hungry. I'm constantly worried about the danger of walking around alone, keeping a wary eye out for danger, but up to now there's been no hint of it beyond the horror stories that everybody tells.

I guess I can tell people that Ukrainians wear the same clothes every day. I assume it's from lack of wardrobe and laundry soap. I see now I brought five times more sets of clothes than I needed to. I probably look too good. And my suitcases are definitely too weighty. It's embarrassing to own so much.

Andrew arrives. He's been in Bulgaria looking into a possible film project and making some kind of a black market deal that was supposed to have yielded a car in Turkey which he could have brought back home to replace his ailing jalopy. Apparently none of that panned out, but he did bring some food and a bottle of wine for the big New Year's party he has planned. That's the day after tomorrow. He wants me to be there. I promise I will.

As for tomorrow, how about a nice little jaunt to the city morgue? This is my idea, not his. He says he'd love to. I've already lined up Ljudmula for this interview, but Andrew says it won't be a good place to take a woman of her age and sensibilities. I agree, and although I'd like to have a good translator for this interview, I ask Andrew to call Ljudmula and convince her to pass this one up.

But she won't. She has assumed total and absolute responsibility for me and will not have me gallivanting around Kiev if she isn't there to protect me. She still doesn't trust Andrew or Volodya. In this case, however, Volodya is not a threat. He has gone to visit his wife and won't be back until after New Year's. So the three of us take a couple of subways and buses through a day of sleet and slush.

The morgue, it turns out, is right across the street from the infamous death pit of Babi Yar. During World War II the Nazis shot over thirty thousand people, most of them Jews, and threw them into this pit. Now there's a giant, horrifying statue in the middle of the pit and a morgue across the street.

This is my first morgue, and I guess half the reason I wanted to go was for the experience. Inside it's as cold as a morgue, and in the hall near the door is a light blue cadaver on a gurney. His face looks aghast, surprised, and one stiff arm reaches out at an odd angle, the fingers spread like those of Michelanglo's Adam about to be touched by God. Ljudmula turns away. She says she'll help with the interview, but she doesn't want to see dead people. I tell her she's quite the trouper and that it's all right with me if she wants to wait outside. She says no, she'll do what she's come to do.

The director of the morgue, a short dark man with a white smock, thick, black-frame glasses and a stigmatism in his right eye, gives me some juicy tidbits. His voice assumes boom proportions as it echoes off the high, tiled walls.

The morgue has never been asked to identify or look out for diseases that might be related to radiation, he says. They've never had a cause of death diagnosed as radiation illness. He has no statistics on the death rate but can certify that the rate of autopsies has increased. Autopsies are performed on anyone who dies while not under the care of a doctor. Until 1987 they received about four thousand of these cases of "instant death" each year. In 1991 there were six thousand, a startling increase of fifty percent.

An even more startling fact is that the deaths have been among the young. Especially common are blood diseases, heart disease, and brain hemorrhage. The latter, which never ocurred among the young, has become commonplace. The cause seems to be thin blood vessels.

Radiation, he feels, may well have something to do

with this, but he says we must also take into account a high rate of alcoholism, especially among liquidators. During the emergency and clean-up, vodka was dispensed, supposedly to ward off the ill effects of radiation, though it makes more sense to me that it was used to bolster confidence and mollify fear. The morgue director says there is some basis to the vodka-as-vaccine theory, in that alcohol robs the body of the oxygen that is ionized by radiation. But he's sure the alcohol does far more damage than the ionization it might avoid. At any rate, many of the liquidators still drink a lot in the vain hope that it might cure some of what ails them or because they fear they might die at any time. Many of them are on pension so, being home and unoccupied, they tend to drink. Since alcoholism is a well-known cause of heart disease, it could account for at least some of the increase in "instant deaths."

I ask him if death by alcoholism might qualify the deceased to be called victims of Chernobyl. He tilts his squarish head to one side and says that it just depends on how you define a victim.

Elena Shezhina, the author's interpreter, and Alex Kuzma, project coordinator of Children of Chornobyl Relief Fund.

Academician Dmytri H. Grodzinsky, one of Ukraine's foremost experts in the field of radiation.

Yuri Shcherbak, Ukrainian Minister of the Environment in 1991, says that all kinds of pollution, chemical and radioactive, in the air, water and food chain, are among Ukraine's biggest problems.

Valentina Patushina and others in the Ukrainian Children of Chernobyl have been keeping handwritten health records of children who were in Pripyat on the day of the accident.

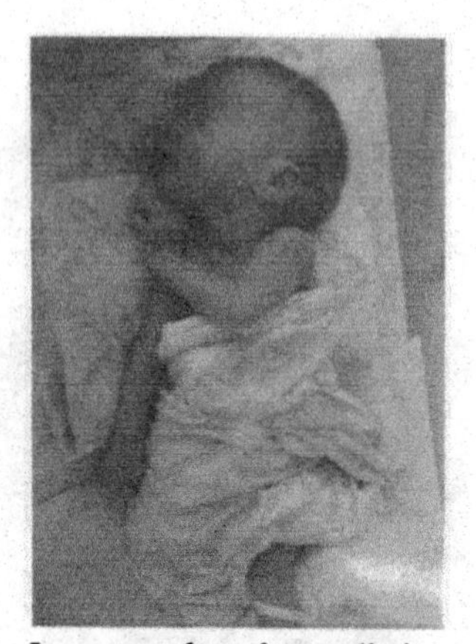

Long-term low-dose radiation is alleged to have caused an increase in the number of underweight babies.

Lelechenky, a Ukrainian children's dance group performing at the International Chernobyl Union conference.

The cancer ward at a pediatric hospital.

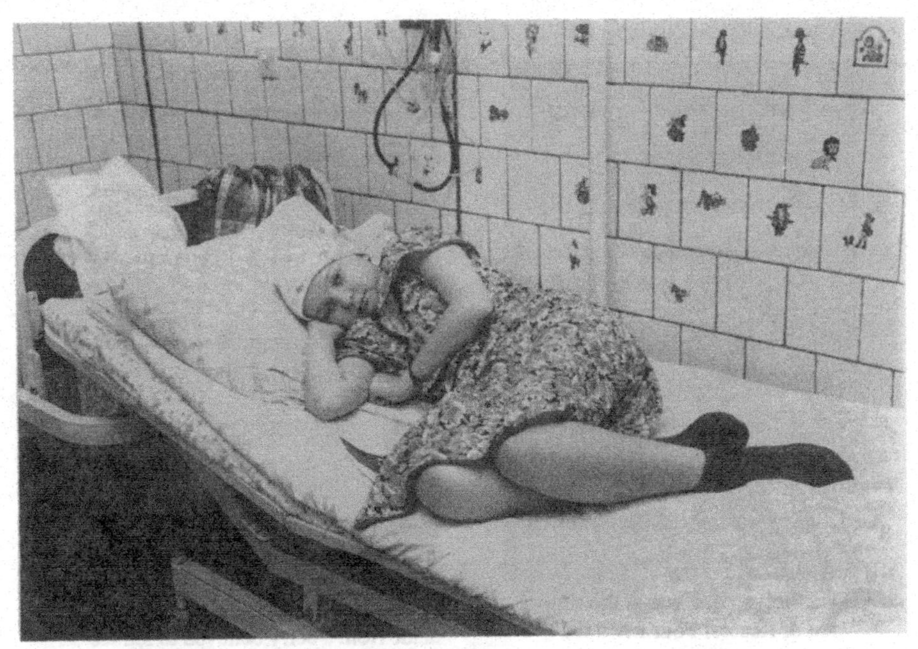

A leukemia victim, not expected to survive.

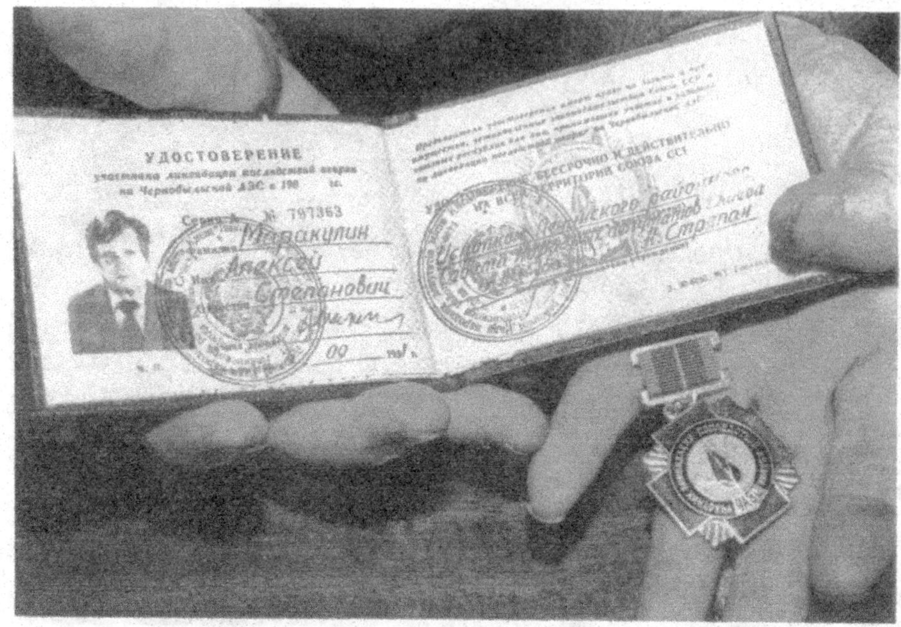

A "liquidator" shows the document attesting to his work to help clean up the radioactive mess, and the medal that all liquidators were given.

Boris Stolyarchuk was at the control board of Reactor Four when the experiment went awry.

Ida Geifman shows two jars of jam made from a berry believed to reduce internal radiation.

Two residents of a village in the Prohibited Zone complain that life was better under Stalin and Brezhnev.

"Old Brezhnev" on the train from St. Petersburg to Kiev.

Elderly residents waiting for the bread truck at a village inside the Prohibited Zone.

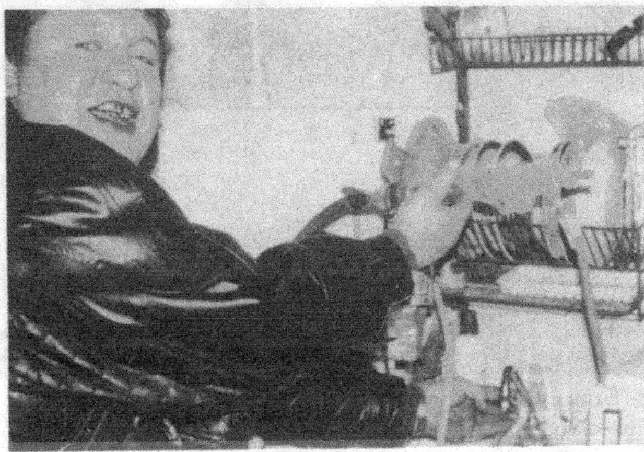

Volodya shows off the distilling device used to make vodka.

Victor's in-laws in their home inside the Prohibited Zone.

The mother of the farmer who lived in the recently contaminated area just outside the Prohibited Zone.

This farmer and his mother live just outside the Prohibited Zone. Their collective farm was shut down in 1990 when radiation levels had risen too high.

The perimeter of the Prohibited Zone.

Houses in an inhabited village inside the Prohibited Zone are either falling apart or being dismantled for use as firewood.

The Dneiper River, which drains the Chernobyl area and runs through the middle of Kiev, is one of the most polluted rivers in the world. Thirty million people use its water. A large amount of radioactive sediment has built up above a dam upstream from the city.

CHAPTER TEN

Pripyat

The workers at Chernobyl used to live in a city built just for them. In a country plagued by shortages of housing and food, a job at Chernobyl and a life in Pripyat were as good as Soviet life could get. Forty-six thousand people lived there.

Today, Pripyat is a ghost town. Its western outskirts are just a couple of miles from the ruins of Reactor Four. The people who evacuated the town were, for a while, pretty much abandoned by the government, but eventually an apartment complex was built for them in Kiev. Ida, the mother of the Ukrainian who put me in touch with Ljudmula, sets up an appointment for me to talk to some people there. Elena will translate, and I am to meet her at a particular subway station.

Which I do with virtually no problem. I get off at the wrong station, but figure that out pretty quickly and find my way to the right place, and there she is. She has a present for me, a nice hardcover book, *Russian Gothic Tales of the 19th Century*.

We go to the apartment of Valentina Patushina. As with most apartment buildings in Kiev, the exterior, the halls and the stairways are bare concrete. The doors are padded on the outside.

Inside, Valentina's apartment is small and stark but pleasant and warm, with colorful tapestries on the wall. While we talk, her mother prepares tea. Valentina tells me how delighted she is that I have come to hear her story. She knows of no other journalist actually talking to people about what happened to them. They feel forgotten by the world and by God.

Valentina and her husband and three children moved to Pripyat to make their fortune, as it were, while Chernobyl was being built. She taught biology at a high school. He had taught physics until he took a job at the plant, where his salary was much higher. He dealt with the installation of equipment and was added to the crew because the project was behind schedule. Everything was being done in a big hurry because of construction problems. To stay on schedule, corners were cut and construction proceeded despite defective materials. Slogans urged everyone to work faster.

On the night of April 25, the weather was beautiful. After the long, cold Ukrainian winter, spring winds were finally blowing in. The big May Day celebrations were a week away. Everybody's windows were open to catch the first breezes of spring.

At 1:23 a.m. on April 26, Valentina heard an explosion like a sonic boom. She had had a premonition of trouble, but it did not occur to her at that moment. Her husband, Sergei, wasn't there because he had gone to the other side of town to help some relatives plant their garden. Valentina went back to sleep, her window still open to catch the radioactive graphite dust and the isotopes of xenon, krypton, iodine, tellurium, cesium, plutonium, zirconium, uranium, ruthenium, strontium, barium, curium, neodynium, neptunium, cerium, lanthanum, and niobium that floated gently on the zephyrs of spring.

In the morning, smoke shrouded the power plant some two kilometers away, and a foot-deep layer of slimey green foam lay in the streets. She had seen that foam before. It was used to wash the pavement when the plant released radioactive materials into the air. It happened often. Sometimes the asphalt had to be ripped up and taken away for burial or was just covered with a layer of concrete. These releases were routine, and the authorities had assured everyone that such minor doses were not dangerous.

That morning, Saturday, Sergei waded home through the stuff and told Valentina that something terrible had happened at the plant. But the radio said noth ing about it, so Valentina and her children, and all the other children in town left for school, wading through the radioactive foam.

Later that morning the city government sent word that no children should be allowed outdoors. There was no danger, but this precaution should be taken. The radio announcements offered no explanation about the nature of the non-danger or how the children should get home. A little after noon some doctors arrived and gave each child an iodine pill — a good idea twelve hours too late. An order was given not to panic.

No one was supposed to go outdoors, but no one knew why not or how anybody was going to get home. That didn't really make sense to anyone. The news on the radio said that all preparations for the May Day celebrations, including races and outdoor practices, would go on as scheduled. The second relay of kids was arriving at school as normal. Spring fever was raging. Kids were leaking out of the school like bees from a hive. The teachers told everyone to go straight home, but it was impossible to believe that the warm sun and blue sky and gusty breeze were deadly. Kids stopped to splash in the radioactive foam, to play in the radioactive sand, to romp in the radioactive grass.

They also stopped to look at the intensely radioactive helicopters that were landing here and there to discharge soldiers and firemen who were vomiting or unconscious. The helicopters blew up clouds of radioactive dust as they whisked off in the direction of the plant where the children's fathers worked. The plant was spewing smoke, and helicopters were flying into it, releasing bags of sand, and flying back to land on the banks of the Pripyat River.

That night Valentina and her family could see the neon-red pillar of fire stabbing up out of the destroyed reactor. In Valentina's words, it looked like it was shining all the way to heaven. Sergei deduced that it had to be an atomic fire because anything flammable would have burned up by then. He told his wife that they would die if they didn't leave Pripyat immediately.

While Valentina packed, Sergei started phoning anyone who might have information. No one knew anything, but rumors were rife. The radio newscasters said that there had been a "trifling" accident at the plant; the result was nothing to be concerned about, but everyone should stay indoors, and no one would be allowed to leave or enter the city. Panic was forbidden.

Unbeknownst to anyone in town, eleven hundred city buses were being driven up from Kiev and were parking along the highway outside of town.

The next morning there were rumors of imminent evacuation. Sergei set off to walk through new piles of foam to warn some relatives who didn't have a radio or phone. By the time he got home, his legs were brown; they looked tanned. He felt dizzy, and soon everyone else did, too. He called a friend who held a high position in the Communist party. He said there might be an evacuation but they would only be able to take documents. Sergei asked how long they would be gone. His friend said, "Forever."

At two o'clock on Sunday afternoon an announcement on the radio admitted to an accident that would require the evacuation of the town. All citizens would be taken to a certain resort town for the next three days. They should bring casual clothes, canned food and documents. No pets could go. Buses would arrive shortly. Everyone should pack and be outside on the sidewalk by three o'clock.

Outside, everyone was dizzy, weak and vomiting. No one really knew why, but everyone shared a sense of dread and danger. Their apartment buildings were locked behind them. As they boarded the buses, they kissed their dogs and cats good-bye and also the daddies who were scheduled to work that day.

Valentina began to cry as she described their abandonment of their home and virtually all of their possessions. We sipped tea for a while — it was made of four herbs, one of which would supposedly alleviate the effects of radiation — and then another evacuee, Natasha, ar rived. Cheese and crackers were set out on the coffee table, followed by some cake. Then some olives and a bottle of Stolichnaya.

Natasha had a similar tale to tell. She said that many people were afraid to talk about it or were emotionally too weak, but she felt it was her duty. The world had to know what they had suffered. She didn't understand why more journalists weren't asking the questions I was asking, and why more doctors weren't interested in the health of the people who had suffered the brunt of the radiation.

Natasha's apartment faced the plant. When she heard the several explosions from Reactor Four, she assumed it was a car backfiring in the streets. She wondered what idiot was trying to start a car at one-thirty in the morning. She never imagined it was a nuclear power plant backfiring. She was going to close the window, but the explosions stopped, so she went back to sleep.

The next morning she saw the foam and felt a terrible premonition. As she watched her children trundle off toward school, she had an overpowering urge to call them back, but her husband told her not to be silly.

But the slimey foam looked odd to him, too, and he had seen the smoke at the plant, where he was due to work on Monday. He called his office to see what was going on, but got contradictory reports. He tried to call the block that seemed to be on fire, but he got a busy signal.

When the children came home, they announced that they were supposed to take showers and have all their clothes washed. They didn't know why, but their father could guess. The reason didn't matter, though. There was no water available. All the water in town was being pumped through firehoses aimed at the white-hot core of the crater around Reactor Four.

Though the radio and everybody official was denying that anything out of the ordinary had happened, helicopters were flying all over the place, and cars that said "Chemical Protection Service" were zipping up and down the streets. Everyone in Natasha's family changed clothes and put on long pants, long-sleeved shirts, gloves and hats. The boy, eight years old, played around the house, but the girl, who was twelve, was petrified. She sat in one place, unmoving, for fourteen hours.

When the buses came the next morning, Natasha

and the children had to leave while her husband stayed behind so he could report for work at the plant the next day. Out on the sidewalk, with everybody reeling from the effects of the radiation showering down on them, his little boy wouldn't let go of his legs. Natasha asked him when they would be allowed to come back. He said, "Never." Natasha turned so pale he thought she was going to faint, so he said, "In two days," but then couldn't resist adding, "Or five hundred years." Then they peeled his son's arms from his father's legs and carried him screaming onto the bus.

Eleven hundred buses weren't enough to hold the whole town, so Valentina's family and two other families — a total of fifteen people — piled into two radioactive cars and headed south.

They stopped at a village sixty-five kilometers south of Pripyat, but authorities there wouldn't let them stay because the village was already overcrowded with refugees. At other villages they stopped to ask the local civil defense people to check them for radiation, but they were always refused. The families continued south, passing Kiev and going three hundred kilometers further to a village where relatives lived. There doctors confirmed that they and everything they were wearing and carrying were radioactive. In fact, they were contaminating everything they touched. Their relatives had to throw out a lot of rugs and furniture. Eventually they and the others who lived there were hospitalized with radiation sickness.

Three days later Sergei had to return to Chernobyl

to report to work. Although his legs were burned from wading through the foam and he was overwhelmingly fatigued, they put him to work filling sandbags which helicopters were dropping into the ruins of the reactor. Most of the other workers on that first day were nuclear scientists and engineers, shoveling madly as if doing penance in hell. Sergei worked until his back gave out.

The whole family was sent to a hospital in Leningrad. Their radiation levels were still too high, so their clothes were taken away. Their radiation levels were still too high, so they had to shower and be measured without clothes on. They were still too radioactive. When their hair was cut off, they were still radioactive, but there was nothing else to do but put them to bed. The doctors who treated them were completely covered, including hands and face, which frightened the children. Valentina completely broke down when they pushed an IV into her arm. She didn't know it at the time, but the doctors were collecting extra pay because the people from Pripyat were considered hazardous materials. When the doctors approached Natasha's eight-year-old boy, they were completely covered in special gear and refused to touch him. That's when Natasha became terrified.

Valentina's family was diagnosed as having distonia (lack of tone in blood vessels). Valentina and Sergei both suffered from blood clots in their legs. Sergei's skin peeled off. The children were so fatigued they could hardly get out of bed. The family wouldn't have known these diagnoses if it weren't for a kink in the bureaucracy that let them get their hands on hospital papers which were supposed to be destroyed.

From the hospital they were taken to a sanitarium. Their daughter was having terrible allergic reactions. Even a change in temperature or the act of combing her hair could set off a rash of blisters. Sergei's radiation readings were so high his records were destroyed. He spent three months in the hospital, then forty-five days in the sanitarium before being sent back to Chernobyl to help with the liquidation project.

When the family was discharged, they were given a single set of clothing and a handful of rubles, barely enough to buy more clothes for the coming winter. This was all the compensation they got for the loss of literally everything they had owned, from home to hair to health.

For the next several months they wandered from friend to friend to spend a few nights and borrow a little money. It wasn't until late in 1986 that they were given an apartment in Kiev. Valentina found work teaching biology, and their lives began to slowly improve.

Kiev did not receive the refugees with compassion. Though the people of Kiev had taken doses well in excess of legal limits, no one wanted radioactive people in their neighborhoods or in the classrooms with their children. Some people refused to share elevator space with Pripyat people. Kiev children were warned that if they played with Pripyat children, their hair would fall out. Nevertheless, some Kiev children ventured close enough to beat up a child from Pripyat.

Five years later, the family is still suffering health problems. Valentina's legs still hurt as they develop blood clots. Since she's a teacher, she has to stand all day. Sometimes she feels herself black out for a few seconds. When she regains consciousness, people are looking at her strangely. Her daughter is diagnosed as having a complex of diseases. She seems to be allergic to her own hormones. Some of her cells have enlarged. Often her throat swells shut.

Natasha's family suffers similar problems. She herself suffered chronic vision problems for several months after the accident. She said it was as if she were looking through insect eyes, with hundreds of miniscule images before her. The floor seemed to swell up at her. Today her vision is normal, but her daughter has second-stage thyroid problems, pancreatic problems, weak blood vessels, irregular blood pressure, liver problems, bladder problems, gastrointestinal problems. Sometimes her blood vessels dilate so much she acts drunk. She has mysterious pains all over her body, and terrible headaches. After school she collapses into bed and can't get up. The doctors don't know what the problem is.

Natasha has cardiovascular problems and often passes out. She wishes she could get over this because it frightens the children to see their mother fall down unconscious.

Valentina's neighbor was what she described as a "fat bureaucrat." He was the doctor in charge of town safety in Pripyat. He was the one who denied there was any problem immediately after the accident. He kept his children home from school, however, and saw to it that they had iodine pills. Another doctor in the building risked imprisonment by calling everyone and warning them to stay indoors. Natasha says she sees both of these doctors often because they live in the same apartment complex in Kiev. Every time she sees the fat bureaucrat she wishes him dead, and she prays that God will watch over the other doctor.

Some of these facts may not be exactly right. Valentina and Natasha are falling all over each other to tell me everything that happened and all the sickness that struck them and how they are being mistreated by the government. The explanations are by no means chronological or in any other way logical. I have a hard time sifting through what is probably fact and what might be exaggerated under the pressure of restrained hysteria. I keep asking Elena to ask them how they know certain facts, who else was present at the time, where I might find these people, and anything else that will help me confirm their stories. The rounds of vodka don't help any of our abilities to organize information.

Elena, however, isn't drinking. She hates the stuff about as much as I hate beets. And it so happens that a lot of beets — pickled beets just too, too radioactive for my taste — are making their way to my dish. Food being scarce, I feel guilty not eating what they put before me. And of course when Elena declines the vodka, she's all but spitting in the face of hospitality. So we work out one of the best deals I've ever made. When a round gets poured and everyone dumps down her dose, Elena fakes it, then sets her little glass near me on the table. During especially emotional moments in the interview, I slip her my dish of beets. She's fed and I'm happy.

During our talk, Valentina's two youngest children tiptoe into the room to watch Tom & Jerry on TV. For the first time in my life I see children actually laugh out loud at a cartoon. When my kid watches a cartoon, he looks as if he's on barbiturates. These kids giggle and guffaw so hard they can't stay on the couch. Even Elena laughs. When I look surprised, she says, "But it's so *absurd!*"

Valentina, Natasha and a woman I'm to meet the next day have formed an organization called "The Children's Fund," which is somehow part of the Chernobyl Union. The main purpose of the organization is to make sure donated food and foreign aid get fairly distributed to children, not to bureaucrats. They've divided Kiev into districts. When donations arrive, they make sure each district receives what it should.

As representatives of the thousands of children who are suffering the effects of Chernobyl's radiation, these women went to President Gorbachev to complain and to ask for help. Gorbachev told them that Chernobyl had been cleaned up and there were no aftereffects.

The group is also gathering information about specific children with specific health problems. They know of no other concerted effort to record the ailments and the symptoms that the children share. They do not keep these records in a computer but rather in a tattered notebook. In several columns they write down the name of the child, the symptoms, the diagnoses, where the child has lived and the material conditions of the family.

When they talk to evacuees from contaminated areas, they often discover that the people have no idea that their odd problems might be caused by exposure to radiation. But when several mothers sit around reporting the symptoms their children show, everyone starts remembering that their children, too, seem to suffer more bloody noses, dizziness, lingering colds, allergies, sore throats, bronchitis, intestinal problems, stomachaches, fatigue than they used to.

Writing all this down in quick, sloppy notes, I get pretty fatigued myself. After six hours of furious scribbling, my wrist and hand cramp up, and my head just doesn't want any more information pumped into it. The words are like drops of water in a Chinese torture. I manage to keep writing only because I feel as though I'm being entrusted with information that no one else has, that the horror and tragedy are being piled into my hands and it's up to me to do something with it all.

Then a man shows up. He's from Pripyat, too, and he used to work at the Chernobyl plant. Would I like to speak with him? The truth is, no, I'm looking forward to getting outside into some silence and fresh air, and besides, the stories have begun to get repetitious. But there's no way I can say no to these people, not after all the beets and vodka they've given me, not after the tragic stories they've just told me.

So Boris Stolyarchuk takes a seat, and I begin my questions. What was his job at Chernobyl? He was an engineer. Where was he at the time of the accident? He was at the plant. Where at the plant? Well, at Reactor Four. He doesn't say this as much as admit it with what seems to me a slightly embarrassed shake of his head.

What was he doing when the reactor blew up? Elena looks directly and seriously into my eyes when she translates that he was at the control board. He and a handful of other engineers were trying to carry out an experiment.

Everyone in the room is silent, their eyes downcast; no one looks at anyone else. I feel something worse than embarrassment. It's as if I have just opened someone's very private wound and showed it to the world. My heart beats hard. Not knowing what to do, I just ask him what happened. Elena translates the question in about three syllables devoid of emotion.

I already know basically what had happened. The directors of the plant had planned an experiment to see if they would be able to shut the reactor down in the event of a general blackout. They were hoping that the turbines could supply the plant with the electricity it needed to shut the reactor down. The question was whether the turbines would continue to spin long enough after they were disconnected from the reactor that normally powered them. Normally, emergency diesel engines would kick in to power the shutdown. But if they started up during the experiment, the operators wouldn't know if the turbines were capable of doing the job.

The trouble was, if the diesel generators didn't start up, the system would automatically assume there was an emergency and, among other protective steps, would drench the core with cooling water. The obvious solution to that problem was to simply shut off the emergency cooling system and, of course, all the other emergency backup systems.

This brilliant plan was submitted to the nuclear energy agency in Moscow, but was never approved or disapproved. The plant directors decided to go ahead with it anyway. It so happens, however, that the design of this plant, which is called an RBMK, makes it very difficult to control when its power output is low. Due to operator ineptitude, the power level plummeted way too low. To increase the level, the director at Reactor Four gave the order to raise several of the control rods that keep the chain reaction under control. The operator who was supposed to do this refused at first, but when his boss insisted, he did what he was told.

It's hard to imagine a nuclear reactor in a more precarious position. It was poorly designed and shoddily constructed. Its directors knew little about nuclear power and its operators knew little about what could go wrong. All safety systems were shut off. The reactor was in a very unstable condition and an experiment was about to be conducted. If it failed, it might very well prove that a nuclear reactor could indeed blow up under certain conditions.

But things started to go wrong even before the experiment reached a critical stage. With the control rods removed, power began to surge. The operators, suddenly afraid, began to lower all the control rods into the core. But an odd feature in the design of the rods actually caused the power to surge even higher as they entered the core. The power level shot up over a hundred times normal in a couple of seconds. The core got way too hot and probably caused water pipes to break. Water hit the core and exploded into steam as it hit the superheated fuel. The force of the explosion was enough to flip the thousand-ton lid off the reactor like a nickel. It came back down in the same place but tilted like the flue of a chimney. Cold air rushed in and apparently set off at least one and probably several chemical explosions as the chain reaction went wild.

The building rocked with explosions and the ceiling fell in. The people inside had no idea what had happened, let alone why. They assumed it was just a steam explosion caused by an overheated pipe — a bad accident but not unholy. Destruction of the reactor was impossible. Of this they were certain. It had seven levels of back-up systems. Everybody in the control room was running around trying to figure things out. They assumed that they were all vomiting because of the tension or perhaps a chemical released in the air. They assumed that the radiation readings on the control board were wrong because they all showed the needles tipped hard against the maximum end — a level that could not possibly occur. They wanted to check the level with some other radiometric equipment, but it was locked up in a steel cabinet so no one could steal it.

But evidence of the impossible was becoming increasingly undeniable. A shaft of bright red light was stabbing into the sky. The nausea was too overwhelming to be anything but radiation sickness. Two engineers ran to the reactor hall to see if they could lower the control rods by hand, but the reactor hall was in flames. In a matter of seconds they took lethal doses of radiation.

The situation was as unreal as a dream. The roar of steam and the crash of the disintegrating building made it all but impossible to hear anyone talk. They tried wearing gas masks, but that made it even harder to hear. They tried wearing gauze masks, but they made it hard to breathe. The air was filled with the indescribable smell of every isotope in the periodic chart swirling through the building and boiling up toward the sky. No one could find iodine pills, so they drank iodine medicine straight from the bottle.

The senior engineer denied any breach in the reactor. He kept insisting that someone run downstairs to open the valve that would let coolant flow around the core, which he still believed existed. Boris refused to go, but someone else went. That brave soul died trying to send nonexistent water to a nonexistent reactor.

The hopeless delusion that the reactor could be intact reached Moscow as a statement of certain truth. Moscow issued an order that coolant be fed continually into the core while authorities rushed to the scene.

Everyone but the top officials knew that there was no more reactor. As the workers retreated toward more distant parts of the Chernobyl complex, they were hysterical with fear. Wracked by spasmic nausea, they were torn between tears and laughter at the horror of it. The horror they imagined was only a tiny part of the whole reality.

Some of the horror stemmed from the fear that they would be blamed for the accident. Boris Stolyarchuk remembers going over and over his every move, trying to figure out what he might have done wrong. But he had made no mistakes. He had precisely followed each step of the experiment.

He and the others remained at the plant until six a.m., when the next shift of workers crossed the lethally radioactive field of fuel and graphite that had been thrown from the core.

On his return to Pripyat, Stolyarchuk was amazed at how calm things were: life there was going on as normal. At the time he had no family in town, so he reported to an infirmary and asked for a whole-body radiation check. The doctor refused.

Stolyarchuk called his friends and told them what had happened, but they refused to believe him. Such an accident was impossible, they said: as soon as the firemen had the fire under control, they would piece together the nature of the accident. It couldn't be as serious as he feared.

He spent all day Saturday pacing in his room. He was still vomiting uncontrollably. Despite overwhelming fatigue, he couldn't sleep. At five a.m. Sunday, a man from the KGB knocked on his door and invited him to visit the KGB office to talk with an expert in nuclear reactors who wanted some information about the accident. It wasn't an invitation he could refuse.

The expert, however, didn't seem to know much about nuclear reactors. His questions were pretty stupid. He kept asking what Boris had done that wasn't in the experiment plan, who had made a mistake, who might be blamed if it wasn't he himself. Boris quickly figured out that he was talking to a KGB agent who was already looking for a scapegoat. Boris stayed with his story, which was the truth — that he didn't know of any mistakes made. The interrogation went on until Boris was vomiting too hard to talk. The interrogator asked why he hadn't gone to the medical center like everyone else. Boris said he hadn't known he was supposed to.

At the medical center, he met everyone from his shift and the ones who had come on duty on Saturday morning. All night they could hear the roar of steam more than a mile away at the plant. No one panicked, although they knew that tons of radionuclides and radioactive graphite were blowing around the air. Boris was expecting a true atomic explosion at any moment. Everybody in the clinic spent the night vomiting in the howl of distant steam.

At nine o'clock the next morning, a bus arrived to take them to Kiev. It had lead screens over the windows. They weren't sure if it was to keep the radiation in or out. From Kiev they took a plane to Moscow, where they were delivered to Clinic Six.

But they weren't let in. They were too radioactive. No one knew what to do with them. Still wearing the light clothing they'd had on in Kiev, they had to stand outside in the cold Moscow air while doctors reviewed the procedures that had been established for nuclear war.

They were put in rooms with three beds each and were free to walk around until it was found that they were contaminating the rooms and nearby patients. As they grew sicker, they were moved to individual rooms.

On the third day a KGB agent showed up and began another intense interrogation, repeating the same questions over and over, asking for repetitions of explanations of the event. Boris was still not sure he hadn't done anything wrong, but he couldn't remember making any mistakes. By this time he was more concerned with not forgetting each detail of his explanations. If he ever contradicted anything he had said before, he would be assumed guilty. He felt he was walking a fine line between witness for the prosecution and for the defense, with himself the one most likely to suffer in the end.

The interrogations continued for the two months he was in the hospital. During that time he was told nothing about what had happened to his friends and colleagues, but by slipping cigarettes to the soldiers who washed the floor, he was able to keep track of who was dying.

From Clinic Six he was taken to a sanitarium near Leningrad, where the daily interrogations continued. He thinks the stress of the interrogations was worse than the radiation.

As his health grew neither better nor worse, he was released, given fifty rubles and allowed to spend a couple of nights in a boarding house in Kiev. Then he was off on his own. He has been on pension since then. Eventually he found a wife, a liquidator who had taken quite a bit of radiation. They have a daughter who was born normal but is often sick, as are her parents.

And that's his story. No one has spoken during the telling of it, and when he is done, no one seems to know what to say. He makes motions toward leaving, but I have one more question. It isn't one I necessarily want an answer to. I just want it asked. I ask when he knew he hadn't made the reactor blow up. Before Elena puts it into Ukrainian she whispers, "Yes, good question." Boris smiles with a kind of humble relief. In fact everyone looks relieved. He says it was in the hospital in Moscow, about two weeks after the explosion. Though the KGB hadn't admitted his innocence, they hadn't blamed him for anything either. The mistake, whatever it was, hadn't been his or any of his colleagues. They did exactly what they'd been told to do, and the whole thing blew up.

CHAPTER ELEVEN

Pioneers

Elena and I go back to Valentina's the next afternoon — the day before New Year's — to meet with Sergei, Valentina's husband. As at our previous meeting, we're served a broad array of foods. I feel guilty with every bite, but I can't keep from reaching for more. While I've gotten used to going all day without food, when it comes along, I partake like a camel tanking up on water. This time Elena drinks her share of vodka, grimacing and shedding a tear as she forces it down. She says it will help her ward off the cold she feels coming on. Same goes for a cigarette, which she smokes in medicinal puffs, quite against her will.

Sergei has just returned from Chernobyl, where he, like eighteen thousand other people, works for two weeks, then returns home for two weeks. During his shift, he and the others who run Reactor One (and, until a couple of months ago, Reactor Two) live in a village just outside of the Prohibited Zone. It's not a bad job, and it could even be said that these workers live safer lives than people in Kiev. They're served meat at three meals a day, and all food is carefully monitored for radiation. They also get twice the pay of their peers.

Sergei is extremely angry about the Chernobyl incident and all that has followed. The reasons for his anger are much worse than any I've heard. He claims that the people who lived in Pripyat at the time of the accident and who now continue to work at the plant have received an accumulated total of 3,000 to 4,000 rad — far more than commonly recognized as a lethal dose. They're finding that children from Pripyat have had as much as 1,000 rad to their thyroids. It has been proven, he says, that children from Pripyat will live ten years less than the national average.

The explosion is not the only cause of their ill health. They are ill also because radiation was routinely leaked for years before the accident. Since there was no way for the radioactive steam to be kept under control, it was simply released every month or so. And there were many smaller accidents that were covered up. The fire at Reactor Two in October was proof that even though the plant was under the most careful scrutiny, a cable could still corrode for years without being noticed. Other cables are in the same condition; another accident could happen at any moment. The plant should be shut down completely immediately, Sergei says. The plan to do so in 1993 is hogwash. Volodymyr Yavorivsky, chairman of the Parliamentary Commission on Chernobyl, has good intentions, but was convinced, by people whose interest lies in nuclear power, that it takes years to turn off such a powerful plant. The truth, according to Sergei, is that it could be shut down immediately — except that doing so would throw eighteen thousand people out of work. That would certainly not be good, but Sergei believes it has to happen sooner or later, and the sooner the better. The total salaries of all those people is forty million rubles per month. In a given month, the plant produces electricity worth only fifteen million rubles. At this rate, the plant will gradually digest all the money in Ukraine.

As for constructing a new sarcophagus, Sergei says it is not necessary, not right now. It is not disintegrating. The alleged holes in the roof, which are supposed to total 1,500 square meters, were built in as part of the structure, to allow ventilation. It would also be pointless to dig up all the buried waste and radioactive rubble so that it won't contaminate groundwater. Such a project would cost billions that would be better spent treating the millions of people who are sick from the radiation.

Sergei has nothing good to say about any government or organization. He says that Ukraine's newly formed Chernobyl Commission is useless, just another bureaucratic machine, a feeding trough for swine. The IAEA investigation was a farce orchestrated by the nuclear powers that financed it. The call for a new sarcophagus is just a ploy to bring in a billion dollars in foreign aid. The Western governments should be paying Ukraine for serving as a massive biological and sociological experiment and for showing the world how dangerous a nuclear power plant can be.

He claims the deception is deeper than we know. All the Geiger counters in the country have been adjusted to read low; he knows because he has checked several against equipment at the plant. From certain sources he knows that the radiation maps are inaccurate. Even if they weren't, they'd be useless because a hot spot identified in red indicates only that a given place radiates over forty curies per square kilometer. It doesn't tell you how much over forty. He says some places have 300. And people still live in these areas. The government doesn't tell them how dangerous it is. In fact, it sends them food and money so they will stay and produce crops and cattle for the cities. If I don't believe it, Sergei says, I can go to the gate at the perimeter of the Prohibited Zone and see crops growing right there.

As Elena and I walk through a cold wind to the apartment of another former resident of Pripyat, I ask her if she thinks Sergei was entirely scientific in his assessment of the post-Chernobyl situation. She says that motives can distort truth. I ask her how I'm going to tell cold truth from the truth tainted by hysteria, avarice, fear and all the other motives that people have. Whom am I to believe? Should I believe the doctors and scientists of the IAEA who took soil and water samples, measured whole-body radiation in hundreds of people, and analyzed tons of statistical data? Or should I believe the mothers who don't know an isotope from a hole in the ground but know damned well there's something wrong with their children? Should I believe a Minister of Health who says there is no problem, or the Minister of the Environment who says there is a problem? Between coughs and sniffles, Elena says she doesn't know.

Our next interviewee is Valentina Rogova, who is assistant chairman of the Children's Fund. Elegantly dressed and reserved in demeanor, she seems the calmest and most precise of the Pripyat people I've talked with. She and her husband had both worked at Chernobyl. Since they had to return to work immediately after the accident, they and other workers were taken to a Young Pioneer camp within commuting distance of the plant. She thinks some twenty thousand people were there and at other camps. They called themselves partisans because, like guerrilla fighters, they slept in tents and underground in bomb shelters. It was hard to sleep at night because people who had been working on the sarcophagus all day spent the night trying to cough the radionuclides from their lungs.

At the plant, their only protection was a cotton mask that covered the nose and mouth. Most people didn't bother wearing it. Mrs Rogova saw soldiers, just boys, standing guard within a few yards of the sarcophagus, their face masks usually dangling around their necks while they smoked cigarettes. They had no idea of the danger they were in. They were bivouacked in tents not far away, where they slept fully exposed to the radioactive dust in the air.

Everyone worked twelve hours, then returned to

the camp for twelve hours. Some people had the unfortunate job of carrying nuclear fuel from Reactor Three because it was feared Reactor Four would explode if the core collapsed in on itself and reached critical mass. The fuel was piled on stretchers and run to trucks which took it away for burial. Other people had the job of gathering up chunks of graphite and fuel. This was a job for robotic machines, but the robots' electronics broke down under the high radioactive conditions. So the people doing the robots' jobs were called "biological robots." Sometimes they took in so much radiation that their muscles separated from their bones like overcooked meat. Doctors were forbidden to tell them why.

Mrs Rogova says that the Pioneer camp was her first experience with real communism. Everyone there was equally underpaid, underfed, unfree and unhealthy. The heroes among them, she says, were those who knew the danger but went to work anyway. She assumes that many of them are dead today and that all the others are sick. She herself has a swollen thyroid and low red and white blood cell counts.

When Mrs Rogova leaves to prepare some food, I talk with her son, a young teenager. He is shy and doesn't have much to say. He was scared at the time of the accident and felt sick. He still feels sick often, but he's studying karate and judo now. He started after some kids in Kiev beat him up for being radioactive.

By the time Elena and I head back toward downtown Kiev, she's feverish and so drowsy she can't keep her eyes open. As we sway in a corner of the subway car, she starts nodding off. When I put my arms around her and pull her against me, she wiggles her face into an opening in my coat and coughs gently against my chest. Her shoulder blades shudder under my hands until we reach the place that sounds like Oh Baloney. I kiss the top of her head, and she steps through the doors just a second before they slide shut. She stands there looking in through the window, but her eyes don't move as the train pulls away.

CHAPTER TWELVE

Happy New Year

The big annual celebration in Ukraine is New Year's Eve. Andrew and Ljudmula have a tug-of-war over who will get to host me for the all-night party. Andrew wins, with Ljudmula consoled by my promise to attend a small dinner party before I leave town.

Leaving town is something I am beginning to worry about. I have an Aeroflot ticket from St Petersburg back to the real world on January twelfth. From what I've heard, it is almost impossible to change that date. Just to try, I would have to go to the Aeroflot office in St Petersburg. Since Russia is now a foreign country and Aeroflot still the Edsel of airlines, the office in Kiev can do nothing to help me. So I have to be on a train to St Petersburg by January tenth. I don't want to leave sooner because I have nowhere to stay in St Petersburg and certainly don't want to spend a hundred dollars a night for an Intourist hotel, not when the Environmental Protection Society hotel is costing me two cents a night.

So leaving town involves a certain precision in timing, and I see a monstrous obstacle hulking between me and home: I still have no visa. If Ana Isabel has ever sent it, it has never arrived. She has no phone, so contacting her is a complicated matter of calling a Portuguese friend of hers and explaining (in Portuguese) that I need to know the status of things, and then I would have to be at a telephone at the moment when Ana Isabel or her friend call back. Andrew assures me that even in best of times, the visa, if sent, will not arrive within my lifetime. Given the political confusion of the past month, it could well be delayed for several generations, assuming it hasn't been lost, which is the more likely case.

But this is not a problem. He will help me arrange a new visa. It is a simple matter of typing up an official invitation from the Experimental Film Society and taking it to a certain government agency which just happens to be right up the street from my hotel, near the state store. So not to worry.

Not to worry! That's a laugh. I've lived in Brazil, where bureaucracy is a way of life, a life dedicated to getting as little done as possible. Registering one's car involved standing in line to get certain documents, filling them out on a typewriter somewhere else, making copies and having them notarized, filing the copies at the transportation department, paying fees at a bank downtown, returning to the transportation department to proffer proof of payment, then waiting while a clerk got back from his two-hour lunch break only to find out that a typo in my great-grandmother's high school diploma rendered the whole process invalid. What would it be like in the tattered remains of a country that was founded and built on bureaucracy? I could picture myself in line at the government agency, a skeleton long-since deceased, draped with cobwebs, glowing softly with strontium, a yellowed visa application dangling from my finger bones.

It's always been hard for me to have fun at a party, and this preoccupation makes it all the harder. The fact is, I'd much rather stay in my room curled up around Elena's *Gothic Russian Tales of the 19th Century,* and Elena, sipping coffee and cuddling my way into 1992.

Actually, Elena has promised to spend New Year's day with me, so it won't kill me to go celebrate with Andrew, assuming I stay sober, which I suspect isn't going to be easy or even polite. Elena has also warned me to beware of "evil women" who might seek to take advantage of the situation to satisfy a certain curiosity that everyone has about American men.

Andrew picks me up early. On the way to his house, I pay careful attention to each subway and bus we get on, counting stops, trying to memorize the names of stations and streets. I want to be able to get back home alone if conditions at the party deteriorate too far.

We're real early, so while his wife, Ira, works in the kitchen, Andrew and I spend a couple of hours exercising his vocabulary, nibbling on olives and kielbasa, and doing silly magic tricks with his son. We watch an American show on TV, all in English. Of all things, it's about hunting for deer in Mexico, where some entrepreneur has fenced off a vast acreage of desert and stocked it with deer. Americans pay to ride along in a truck with a ten-foot tower on the back. They sit up top and ride around until they flush a deer. Then they shoot it. One hunter holds up the antlers of a buck and drawls, "This rack shows an animal with pride, spirit and character and I am proud to be the one to harvest him." It is a very strange thing to see on New Year's Eve in Kiev.

Following a Ukrainian tradition, Andrew sets small bowls of food on the window sill just behind a curtain. These are for spirits who, if made happy, could ensure a good year. He says he doesn't know if it will work, but there's nothing else to do, no other hope.

Eventually two other couples arrive, and Ira emerges from the kitchen. We all settle around the low, wide coffee table in the center of the living room. Andrew translates the many questions they all have for me until the conversation develops into something closer to normal. Andrew trans-lates only the important comments. To my amazement, I often understand what they're talking about. The occasional word that resembles English — *teknologee, radiatsee, Americansky, beezneez, kamikazee, Gorbachev, Chornobyl* — and the few words I've learned in Russian, are enough to let me guess the gist of the talk. A couple of times I laugh at a funny story a second before it gets to the punch line, and everybody looks at me suspiciously. Andrew would later tell me that one of his friends pulled him aside in the kitchen and warned him that I was not who I pretended to be.

The drift of the humor is of the gallows type, that life in Kiev has become intolerable and is bound to get worse. My presence reminds them to add radiation to their fears of political chaos, famine, unemployment, shortages, economic depression, and absolutely not one sign of anything good coming down the pike except that the Soviet Union is gone and Ukraine is independent once again. They keep apologizing for being so gloomy, but to me they look happier than people at a lot of parties I've been to. They do shots of vodka, each time raising their glasses in toast. They dance to music played on a little cassette deck. I guess I look uncomfortable because I am the only one not dancing. Ira comes to urge me to dance with her while Andrew pretends to take a little nap on the couch.

She's small and light and pretty flexible under my hand at her waist. As we turn clockwise among the other two couples, her milky eyes look at me as if to say nothing less than "take me, kiss me, I love you." She pulls in for a closer dance, her arms reaching around my neck, my arms going all the way around her waist. As we spin by the light switch, she flicks it off. Every few seconds she lifts her head from my chest to give me that milky, moony look. I'm thinking maybe it's a Ukrainian tradition to give your hostess a little kiss behind the ear while her husband isn't looking . . . either that or you're absolutely *not* supposed to even so much as *think* any such thing, especially not about the wife of the man who has been so helpful and giving since about three decades ago when I arrived in Kiev and didn't know a soul save for one Ljudmula, who hadn't answered the only door I had to knock on. So I keep that kiss in my pocket where it belongs. The music comes to an end. The light comes on. I quickly go sit.

The food keeps coming out of the kitchen, everything from smoked squid to fatty salami to fried goose to pickled this and that. We drink Moravian wine, Russian Stolichnaya, Ukrainian *samagunke,* Bulgarian champagne. The room fills with cigarette smoke, and even though I'm responsible for a good part of it, I need fresh air. I feel a bit woozy and even worse, as more and more and more food gets piled onto my plate. Now I know I'm really insulting the hostess. I'm not keeping up with what she's dishing out. The beets are becoming an especially big pile. And whenever everybody bottoms-up another dose of *samagunke,* I palm mine and tuck it behind something. At a moment when everybody happens to be out of the room, I scrape a hefty serving into the dishes set out for the spirits who, so far, haven't made a dent in the offerings. If they like beets, 1992 is going to be a very good year.

The food really is good; there's just too much of it coming down onto too much vodka. And the people are good, the talk good — everything's good, but by four in the morning, I want out. I want fresh air and elbow room. I want to go see Elena, maybe catch a few Z's before she arrives. I'm worried she'll arrive before I do and then leave. When the women fall asleep and the men set up conversation in the kitchen, I start talking about going home. Andrew won't hear of it. The party has just begun; if I go out into the dark, someone will kill me; I have to stay for coffee; it isn't 1992 in America yet — stay until the New Year arrives in your country!

Okay, that's it: I'll stay until midnight E.S.T. We'll drink in the New Year a second time and then I'm on my way.

There's nothing like a slug of *samagunke* at seven a.m. It hits the gut like an ice cold bowling ball. They pour another. I absolutely positively cannot drink it. It will not go down my throat. I beg off but have to compromise by agreeing to have coffee. The women are awakened, coffee brewed. It's good, and I'm ready to go, but then more coffee arrives. It's good too, but then they're piling more beets on my plate, and anchovies, squid tentacles, something akin to potato salad. I can't do it. I am a horrible, ungrateful, ungracious guest, but I cannot eat or drink anything more. I have to go outside. One of Andrew's friends understands, invites me over to his house for some fancy cognac. But that sounds to me like an invitation to an early morning drinking contest fueled by lack of common language. I beg off.

At last the sun comes up. The day is cold. For the first time in four weeks, I see blue sky. Andrew, not believing I can find my way home, escorts me on the bus to the subway, but once there I convince him I can make it on my own. To prove it, I successfully name each and every bus and subway stop between there and home. I can't tell if he's impressed or suspicious. We shake hands and wish each other ten flavors of good new year. Feeling very guilty for bugging out so early into the party, for being so socially inadequate, I thank him profusely for inviting me. The bus arrives. We part, and thank God Almighty I am free at last.

* *

Back home, I take a long and detailed shower, washing the cigarette smoke from my hair and the stagnant puddles of vodka from my brain. Afraid that I won't hear Elena when she arrives, I stay awake all morning, drinking coffee, reading Gothic tales, trying to rest one eye while the other stands watch. The intoxication of sleeplessness reminds me of the effects of marijuana. I feel pleasantly stupid and stupidly pleasant. I have this sick fantasy of lying in bed with Elena while we both read ourselves to sleep. Can I help it? My heart is weakest when I'm almost asleep. Several times I doze off, then snap awake, look out the window, listen down the hall, scrutinize a distant figure shuffling down the sidewalk, but she never shows up.

CHAPTER THIRTEEN

Not to Worry

Andrew knows a former liquidator, Victor, who has relatives living within the Prohibited Zone. They and a hundred other people have remained in a village just fourteen kilometers from the Chernobyl plant. Victor has agreed to take me there. It's slightly illegal, but he can sneak me in if I give him fifty dollars for black market gasoline and the rental of a friend's car.

Something smells fishy here, but the prize is too tempting to pass up. Sneaking into the Prohibited Zone to talk with people who supposedly do not exist is just the kind of journalistic coup I've always wanted to pull off. But is it really possible to simply hop in a car and drive past the KGB to take a tour of a town that was supposedly evacuated six years before? It sounds like the kind of expedition that never gets where it wants to go. Though I'm reasonably certain it will be a fiasco, I agree to try. But I'm not parting with my fifty bucks until we've gotten there and back.

But first, it's time to get cracking on that visa! Obviously it isn't going to arrive from St Petersburg. Ljudmula has promised to phone Ana Isabel's friend to get information, but I'm certainly not going to wait. Time is short. Today is January second. The third is a Friday. Nothing happens on the weekend. On Monday the sixth I must go see Ljudmula for a little party in my honor. On Tuesday we go to the Zone. On Wednesday I take the train to St Petersburg, arriving on Thursday morning. My flight leaves on Friday. If I'm not on it . . . if I'm not on it . . . I literally cannot imagine what will happen.

In other words, today is just about the last possible day to go look for a visa. So I'm pretty crestfallen to find that the government agency right up the street is just an abandoned building with a tattered note on the door explaining where it has moved. But Andrew says not to worry. The other place is not far from Victor, the liquidator who's going to take us to the Zone. So we hoof on over there, a hike of about a mile. Thinking farther ahead than I normally do, I suggest we stop at a photo shop for a set of little pictures that will undoubtedly be called for in the visa. By "photo shop" I mean a place whose only function is to take little black-and-white pictures for documents. They're only a penny each, so I order a dozen. Though they won't be ready until Monday, I insist that we press on toward seeing about a visa. It would sure be nice to have when we go to the Zone, a trip which is bound to feature at least one conversation with a cop.

Victor lives several flights up in an old building. His apartment is really just a bedroom in a communal apartment he, his wife and daughter share with another family. About half his cluttered living space is taken up by a double bed. As a liquidator who legally took in the maximum allowable dose, but in all probability a lot more than that, he now lives on a pension.

Victor was sent to Chernobyl as an ambulance driver on May fifth, a day before the atomic fire was extinguished. He spent a couple of days in the town of Chernobyl. From there he could see the helicopters ferrying sand to the reactor. There would be two in the sky at a time, with a couple more being reloaded down at the Pripyat River. They carried the sand in parachutes slung below, and often they hurried the trip a little too much. Hovering low in the radioactive smoke, they tossed their payload too soon and missed the target.

As soon as the fire was out, Victor was sent in with his ambulance. He had little idea of the danger. He drove around the plant a couple of times looking for the person he was supposed to pick up. Finally someone appeared with an unconscious soldier on a stretcher. Victor drove him back to Kiev.

In September he was sent with a team to help people in a nearby village. The health problems seemed to be related to radioactivity: a woman with gastrointestinal ailments, a soldier experiencing dizziness, things like that. One of Victor's jobs was to visit a kindergarten and explain to the children that they shouldn't play outdoors. While in the classroom, he took a reading with his Geiger counter. It showed six milliroentgens per hour — fifty percent higher than the area in and around Kiev right after the disaster.

In 1987 he was sent back to Chernobyl to drive a tow truck which was collecting all the radioactive vehicles in the zone for burial. His truck was protected by lead shielding, but his friend's truck wasn't. He was also assigned the duty of gathering a certain fibrous material that was stuffed into cracks in the sarcophagus. The material had to be changed daily and buried.

His boss sent him into Pripyat to steal things that the evacuees had had to leave because they were all radioactive. There was a lot of this theft going on. Victor stole some furniture from an apartment, a certain book from a library, and some cement. His boss sold this stuff to buy vodka, and presumably the buyers had no idea the goods were radioactive.

Victor still has his Geiger counter. All it needs is a battery. I agree to buy one before our trip. With the use of a converter, it runs off alternating current, which is not available in the village in the Zone. But he hooks it up to show me how the needle leaps when he holds the receiver near some radioactive soil he has. It's in the bottom section of a can, sealed in with wax. Sure enough, when he waves the wand near it, the needle jumps way up and the machine chatters with static.

Victor has a few souvenirs for me. One is the radiation badge he had to wear but never turned in because he knew they wouldn't tell him the truth about its reading. Another is a photo of himself standing just outside the sarcophagus. Still another is a list of radiation readings taken from food in the Kiev market. He had copied the readings from some documents he was supposed to be transferring between government offices. The readings are very different, he says, from official reports.

Victor goes with us to pursue my visa. It's a long, long walk across town, first to the wrong building, then to the right one. Then we sit and wait until somebody gets back from somewhere. It's a woman, and Andrew suddenly seems worried. When he explains the problem, he holds his fur hat in both hands down near his belt. The woman gets mad. When she barks a question directly at me, I catch the Russian words for "where" and "visa" so I go right ahead and say "Saint Petersburg" with a finesse of pronunciation that surprises even me. Andrew gives me a suspicious look. The woman goes ahead and asks me something else that sounds a lot like "What the *hell* is it doing there?" Andrew hesitates before translating it as "Why?" I explain that it was expired and it was for the Soviet Union, not Ukraine, so what does it matter?

Well, it matters to her. She is very, very miffed that I have entered her sovereign state without a visa. I ask Andrew to tell her I'm a journalist writing a book about Chernobyl, but he says that is a fact that she'd better not know. I wish he would ask her if she has any children and if they've been passing out lately. He translates my thought as something else entirely, something that is good enough to get us out the door without being arrested for espionage.

Great. Swell. This is just dandy. Marvelous. I'm so tickled, so pleased to have waited so long, to have believed the "No problem, no problem" I've heard so many times. I *knew* this was going to happen when I handed my old visa to Ana Isabel, but I went along and bit by bit let things get out of my control, and now I'm going to pay.

But Andrew insists this is not a problem: we will go to the Aeroflot office and change my ticket to a later date. This means another long walk across town to the Aeroflot office. Andrew is pleased to find a short line there, but to me it looks like there are at least thirty people standing there. We wait and wait and wait and then go right up to an unoccupied agent behind the counter. She says she doesn't know if they can do anything about a ticket issued in Russia, and even if they could, there are no open seats out of the country until February 26 — a month and half away.

Isn't that nice. Another seven weeks here when I should be home paying my mortgage, part of which payment comes from the classes I'm supposed to teach at two universities starting in the middle of January; another month and a half in a place where I have simply no idea where to find food, let alone such luxuries as toothpaste and soap.

Hungry, tired, confused, a little scared and not at all happy, I shake Andrew's hand and hear one more "No problem" before he heads for home. Victor boards a city bus with me and we ride for a long, long time, then change to another bus and ride for another long, long time. I start to worry about this. I'm sure we are wandering around parts of Kiev nowhere near where I live. But eventually we arrive at Stadio Centralee and I bid Victor adieu, perhaps in a manner less friendly than he deserves.

My only hope now is that Ljudmula will somehow discover that Ana Isabel has taken care of my visa.

But that is not going to happen. Ljudmula comes to see me the next day. With her is a journalist, Nikolai, who works for a newspaper called *Voice of Chernobyl*, which deals exclusively with the Chernobyl issue. Before we start to interview him, Ljudmula reports that she has finally managed to receive indirect word, from someone who phoned someone who talked with Ana Isabel, who phoned someone who phoned Ljudmula. The visa is underway somewhere but still has not been issued.

But not to worry.

I can't not worry. The repercussions of not having a visa are too far-reaching and serious. And complicated. How am I going to live here for several more weeks? How am I going to arrange things back home? It's all but impossible even to phone there. Overseas phone calls out of Ukraine have to be requested well in advance. The operator will tell you, for example, that your call will be put through at three o'clock next Thursday morning, and you get just a few minutes, and that's only if the connection doesn't fail, which it is most likely to do. The more I think about the complications, the worse they look, and the worse they look, the less I can concentrate on the business at hand.

The business at hand is a journalist who has been inside the sarcophagus. He and a photographer went in for a firsthand glimpse of the conditions under which a certain doomed group works. Since they have already taken in doses in excess of two hundred rad, they have resigned themselves to early deaths. It makes more sense than exposing more people to deadly doses. For the last four years, these workers were compensated with salaries five times higher than those of comparable workers in other industries. Now they're earning just double the standard wage, which is the same as other Cherobyl workers who work under less radioactive conditions. Many of the sarcophagus workers, taking this as a slap in the face, have quit. Others aren't happy, but stay on in the name of humanity.

What they're doing is gradually searching for and mapping the location of the nuclear fuel that was blown into the hundreds of nooks, crannies, closets and rooms of the reactor building. With radiation at varying intensities throughout the building, and rubble blocking many places, workers can't just walk around looking for the stuff. They have to drill long holes through several walls and insert electronic periscopes to take readings from which quantities of fuel can be estimated. The drilling is technically difficult and the estimations are less than perfectly accurate. Since no one knows for sure how much fuel was blown out of the building — I've heard everything from three to eighty percent — they can never be sure if they have located all the fuel that remains.

Making matters worse, the fuel keeps moving as gravity and rainwater do what they do to everything God and man have built. The atomic fire after the explosion was hot enough to melt concrete. After the fire, it became radioactive dust, which has mixed with the radioactive fuel and radioactive graphite in which it was packed.

Nikolai was there on June 24, 1990, when enough of the fuel trickled into one place to form a critical mass. Fortunately, some workers were willing to commit suicide, albeit not instantly, as they desperately searched for, and found, the concentrated mass.

Nikolai did something dumb — the kind of dumb thing I'd probably do, too. He took a tour of the inside of the sarcophagus. It was not a casual stroll. At certain points, he and his photographer had to run through especially radioactive areas — dashing down a hallway, for example, then going through a door and down a ladder within so many seconds. They went one at a time so there would be no possibility of their tripping over each other.

At the innermost point, there was a thick window that looked down on the rubble of the ruined reactor. One could look in if one wanted to, but a two-second glimpse would "cost" 2.5 rad — half the allowable annual dose for a nuclear worker, or five rad a year. Because radiation will go through glass.

Nikolai had to do it. He stuck his face to the window, gawked for two seconds, and pulled back. All he saw, of course, was a pile of blackened brick, wire and debris.

But that wasn't the point. The point was that it was there to be seen and so it *had* to be seen. It scared me just to think about it because I knew I would have looked, too.

The radiation ruined all but half a dozen of the couple hundred black-and-white photographs Nikolai and the photographer took. One was of the clock that stopped at exactly 1:23:46 a.m. on April 26, 1986. It reminded me of the picture of the tower clock at Hiroshima that stopped at 8:15 on the morning of August 6, 1945. The other pictures were of workers in cloth masks who worked in there day after day.

Nikolai had one other interesting picture that was taken after the Reactor Two generator caught fire in 1991. The shot is up through the fallen roof, a hole roughly fifty meters square. In the center of the sky is a blurry picture of what could well be a flying saucer. Nikolai said he doesn't know what it was and that the photographer didn't see it when he took the picture, but the Soviet Ministry of Internal Affairs, the equivalent of the US Federal Bureau of Investigation, analyzed the picture and verified that it had not been faked. But they didn't know what it was either.

CHAPTER FOURTEEN

The Author as Swine

I owe Ljudmula my presence at a dinner party at her apartment. It's the least I can do — out of sheer hatred of the Soviets, she refuses to take any payment for all her interpreting — but I turn out to be a less than rewarding guest. Her son picks me up in his car, using the last few liters of gasoline in the independent sovereignty of Ukraine. As a gesture of thanks, I bring all the food, toothpaste, toilet paper, writing paper, pens, pencils, and other supplies I figure I won't need. I also bring a nice new sweater for her to pass along to Elena.

When I arrive, Ljudmula confirms my worst fears. Ana Isabel hasn't made any progress on my visa. In fact, the old visa is now in a locked drawer of a desk in a locked room at the university, which is closed for the holidays. Her message is not to worry; she'll think of something. Ljudmula appends to the message a reminder that I have a tendency to entrust myself to the wrong people.

Literally worried sick over my visa problem, I am an absolute pill during the whole soiree. While two other women fuss in the kitchen, rustling up a meal that probably cost days of waiting in lines, Ljudmula, ever the trouper, translates a long article written by Nikolai. She reads the whole thing into my tape recorder, not knowing until we're done that the batteries are failing, and are converting her voice into an unintelligible moan.

Somehow Ljudmula hasn't learned that the next day I am going to the Prohibited Zone with Andrew and Victor in their borrowed car. On her own initiative, she has asked someone to arrange an interview with Volodymyr Yavorivsky, chairman of the Parliamentary Commission on Chernobyl. To my surprise, the arrangements are made and the confirmation arrives by phone while I'm sitting there. Ljudmula, rightfully expecting me to beam with delight, is shocked to hear of my other plans. The person calling is also shocked to hear that his efforts to line up this interview have been in vain. During this tumultuous time in Ukraine's three-week history, Yavorivsky has agreed to reschedule his day and arrive at parliament late in order to talk with me. It is unthinkable that I could now decline.

But it's also unthinkable that I could miss this chance to visit the Zone. It is my only chance to see Ukraine outside of Kiev, and it is big-time adventure to be doing something risky in pursuit of the truth.

My decision hurts Ljudmula. I have devalued her efforts to help. I have been doing things behind her back. I have embarrassed her before whomever set up the interview, and worse, I have embarrassed her before a government official. Perhaps worst of all, I have rejected her in favor of Andrew. I didn't even invite her to come with us. I didn't offer to hire her son and his car to take us to this place.

Of course I haven't *meant* to do any of these things. I thought I *had* told her I was going. I didn't invite her because it didn't seem like a suitable trip for a woman of her age, not to mention that she probably wouldn't fit in the car. And I hadn't known she was lining up another interview for that same day.

Nothing can be rescheduled because I have a train ticket for the next morning. Ljudmula's friend calls back to try to convince me that it isn't worth the trip to go see a bunch of old people too stupid to move out of a radioactive zone. We call Andrew to see what he has to say, and he says an interview with another government bureaucrat isn't going to yield any new information. Ljudmula and Andrew quietly argue with each other over the phone, and then we sit down to eat.

Mad at everybody, from Ana Isabel to Andrew to Ljudmula and most of all at myself, I can't bring myself to say anything civil at the meal. One of the women present is the chronically cheerful type, the type who thinks people will brighten up if you tell them to, the type who tries to force happiness into people by telling them to smile. But I am not so easily charmed. Clenching *shove it* between my teeth, I crank up a smile that makes Chernobyl look like a bouquet of daisies.

And so goes the evening. I treat Ljudmula's friends the way zoo animals treat their visitors. I eat her hard-won food as if the world owes it to me. I behave as though she is a mother I have every right to abuse. And then I leave.

CHAPTER FIFTEEN

Prohibited Zone

Andrew and Victor — the ambulance driver who had gone to Chernobyl shortly after the accident — and Victor's wife pick me up well before dawn. Andrew and I sit in back, nestled in our bulky coats. As we drive out of Kiev, we throw seat belts over our shoulders when we pass police checkpoints, and Victor's wife puts a babushka over her head. We don't look out the windows. We just drive by at slightly slower than normal speed. Then the seat belts and babushka come off.

An hour outside of town, we stop at a village that has a grocery and a restaurant-like entity. By gosh, the grocery has bread and sausage, the restaurant has soup. Andrew and I have some soup. Studded with beef by-products, it's pretty good. Victor buys a bag of food. Everybody gets happy. Life looks good.

I keep looking out the window for signs of radiation in the fields of shallow snow. Everything looks like normal winter's death. At a village just outside of the zone, Andrew asks if I'd like to stop and talk to anyone. I would. We park in front of a little house with a barking dog. I guess this is a typical little Ukrainian farmhouse, with white trim and designs painted on the light green walls. Several colorful cooking pots are drying on a tall post that has pegs sticking out of it.

Andrew knocks on the door. A little farmer in rubber boots and black-frame glasses comes out. I have to show him my passport to prove I'm foreign. He agrees to talk and invites us inside. His mother and father are there, visiting from somewhere. His mother is built along the classic rounded lines of every other old woman in Ukraine. She's cooking bread in a brick wood-fired oven. His father is geezing around in boots. They let me photograph them.

The farmer explains that he was part of a farm cooperative shut down a couple of months ago because of radiation. After six post-Chernobyl years of the cooperative's producing wheat and milk and potatoes, the authorities found radioactivity there. He's suspicious about this. He thinks the decision was political or economic, but he isn't sure. Once a scientist came and took readings and said nothing was radioactive. But then he went next door and said the neighbor's cow was giving radioactive milk. The farmer doesn't know what to think, but he doesn't want to move. His parents insist he should move because now, with everyone else gone, there's no place to buy milk or bread. But he says he has his house here, and a cow and a pig, and doesn't want to leave.

I don't blame him. He's got a nice little plot of land with a big field behind it. He's sure there's no radiation. He's never seen a sign of it, no two-headed animals or anything. He takes us down the street, a lane of broken asphalt, to some houses that are being dismantled to be moved to wherever the inhabitants have gone. That's proof, he says. Why would they take houses if they were radioactive?

We drive on. Victor informs me we're in the Zone. I wouldn't have known, except for the barbed-wire fence along each side of the road. Still no signs of widespread death, but Victor pulls over so I can take a picture of a little sign that's hanging on the fence. It says "Prohibited Zone." I maneuver into an awkward position so I can have a dead tree in the background. I take half a dozen shots.

We stop at a guard post on the far side of an intersection. It's two stories high, with the business room upstairs behind a wall of windows. A red-striped bar juts over the road. Victor goes into the guardhouse to ask if we can visit the little cemetery where his wife's parents are buried. It's just inside the gate. The guard says okay. Andrew whispers to me not to take my camera out of my bag. Speaking in lowered voices, not looking to either side, we saunter over to the cemetery. It's an old, overgrown plot studded with rusty iron crosses in the snow. It would make a great picture, but Andrew says no. He seems very nervous. Victor and his wife traipse off through the snow to pay their respects at the appropriate cross.

On the way back to the car, Andrew explains the purpose of a little guardhouse off to the side. It's the radiation checkpoint. Any car passing through the gate must stop here for a reading. But such cars are few because normally a car would have to remain outside, its passengers transferring to a car that always stays in the Zone.

Inside the little house, half a dozen young soldiers are watching TV. Andrew explains who I am and asks if I might take a picture of somebody checking a car for radiation. They say no, and Andrew seems to beg forgiveness for having asked.

One of the roads at the intersection leads to the little village where people still live. We leave the car near the gate and walk in, a hike of about half a mile on a raised roadbed that passes between two low fields. I try very hard to feel some neutrons or something, but all I feel is damp cold. Victor points out the distant steam rising from the Chernobyl plant. It's fourteen kilometers (8.6 miles) away. I wonder if this little jaunt is going to give me cancer. Maybe so; probably not. I'll always wonder and never know. That alone may well knock a few years off my life. But wonder I must, and if risk brings me closer to truth, I'll take what I get.

The houses in this village are cabins with squared logs and thatched roofs. I see a few chickens and am told there are cows and pigs present too, although livestock is strictly prohibited. Most of the houses are empty and many are falling down.

The village and the peasants seem right out of a fairy tale, little people with big hearts, the kind who don't let a little strontium interfere with their way of life. Victor's mother- and father-in-law weren't expecting us. We catch them sawing up a log they've pulled from a cabin abandoned by people they knew. With the log at shoulder-height in a sawhorse, the two of them — he in high black boots and fur hat, she in kerchief and heavy coat — pulling on each end of the saw, look like some kind of Arkansas lawn ornament driven by the wind. Between the two of them, they are a century and a half old. When we look over the gate, they drop the saw and make sounds of elderly glee.

This is Zone One. It's a perfect place to grow radioactive cucumbers and steep them in a radioactive brine of radioactive herbs to make radioactive pickles. But these people don't seem too worried about radiation. For one thing, they're too old to have to worry. For another, what else are they to do? Go share a communal apartment in Kiev? They probably couldn't find one even if they wanted to. They'd sooner die, and they seem in no hurry to do that. Everybody in town is seventy or eighty years old. The average height is about five feet, the average eye color blue, the average teeth gleaming gold and silver. It's a beautiful sight to see these people smile in sunshine.

One woman remembers the revolution. She hid in the woods with some children while the fighting passed through. She remembers World War II and Stalin. She spent her life laboring on this cooperative farm, trying to make it work. Now the Communists are gone and the farm is dead. She's sad to see it pass.

Another old woman remembers spending eight months in a distant village after the explosion at Chernobyl. During the first three months they had to live with another family. When they came back, it was a long, cold winter but at least they were home. Her husband worked at a cooperative farm and got food there, but there was no help from the government. He got a thousand pounds of potatoes, most of which they exchanged for a cow and a horse. Not too long ago they sold the cow to a gypsy. Now they raise their own food and send some of it to her daughter in Kiev. She says she will stay in her village until she dies.

We walk over to a house where a relatively young man — sixty, maybe — lives. He works at Chernobyl. While we talk, his chickens run around and his cat rubs up against my legs. His dog scratches at fleas. We all munch on the roasted sunflower seeds his wife gives us.

He remembers the day Chernobyl blew up. Helicopters were flying all over the place like bugs. At first the officials didn't force anyone to leave, but they killed all the animals and chickens. Only on May third, more than a week after the explosion, did they start moving everyone to another village.

This man denies there is any radiation. He says the government comes in and checks everything and says everything is normal. He says once he took crops into Kiev and had them checked and they were normal. They checked him, too, an all-day process that made him feel like an astronaut. They found nothing abnormal. But as we leave, Victor says the man is lying. What could I expect, that the man would say, "I'm an idiot and I live here even though

I know it's going to kill me?"

We talk to a lot of people. They seem eager to tell a reporter that things are bad, foodwise. They don't want to know about cesium isotopes. They want bread. They want prices down where they used to be under good old Stalin and Brezhnev. Radiation is their last concern. I ask how they feel having no children around. They say it's not a problem. It's actually kind of quiet. What they want is bread. The government truck that brings it hasn't come by in two weeks, but what does it matter if the prices are too high to afford?

We have lunch in the little fairy-tale house where the in-laws live. The walls are blue, the ceiling barely above my head. It's warm and steamy from the cook-fire. When I try to take pictures, the lens fogs over. The old woman serves us some excellent potatoes, beef, slaw, beets, and the best pickles I've ever, ever had. Very heavy on garlic and other spices, nice and crunchy. After two or three vodkas, I am praising the pickles as the best in the world. I'm even eating the beets. Fog or no fog, I take pictures of everything, from the pickles to the old man's bald head, a nice sign of radiation if he weren't eighty-two years old. When we leave, the old woman gives me a plastic bag of pickles. Radioactive for sure, I hope. These pickles are going home to Connecticut for show and tell.

Incidentally, I brought these pickles through US customs without any problem. A beagle in a gray vest passed my suitcase through without showing any visible interest in it. Later I took the pickles to the radiation officer at Yale-New Haven Hospital for Analysis. It turned out they were a tad below the background radiation of New Haven, Connecticut.

CHAPTER SIXTEEN

No Exit

We barely make it back to Kiev. The gas tank is all but dry when we pull up to the Environmental Protection Society. This is my last night in Kiev! Tomorrow I take the train back to St Petersburg. Andrew and I embrace, pound each other on the back. I have this terrible feeling I should thrust some money into his pocket, but he has said repeatedly that he does not want to be paid for helping avenge the Ukrainian people for the disaster they have suffered at the hands of the Russians. He's lying, of course. Like everybody else in town, he and Ljudmula and Elena most desperately need money, but it will take me a few weeks to figure that out.

Andrew leaves me the key to his office so I can call him if I have any problems before morning. He has already called the state taxi company to have someone pick me up at six a.m.

I am dismayed to find the shower room locked. I haven't washed in three or four days. The last couple of days I figured I'd wait until after my trip to the Zone so I could wash the radiation from my hair. But now it looks as though I'll be taking it with me to St Petersburg.

I call Elena to say good-bye. She's sorry that she can't see me before I leave, but she says maybe it's better this way. I ask her what she means and she says I know what she means. I guess I do. She keeps asking me not to hang up, so we talk for a long time before we say good-bye.

All night my spine is in knots as I think and rethink of all the things that can go wrong between here and home. Will the taxi arrive on time, and if not, what will I do? Will I be able to figure out which train to get on? Will the train arrive on time in St Petersburg? Will I be able to find Ana Isabel's place? Will they be home? Will they have a visa waiting for me? If they're not home, where will I go? What will I do without a visa? If they have the visa and I make the plane, has the plane been maintained the way Chernobyl was maintained? Or has even that kind of maintenance been forgotten since the recent collapse of everything?

All these problems would be a lot easier if I didn't have two suitcases that weigh fifty pounds each, plus a camera bag. I can't carry them more than a few feet without stopping. And I look as ostentatious as the Pope, lugging around more stuff than most Ukrainians own.

If I didn't have this weight, if I had my belongings in a backpack as I usually do when I travel, I'd feel perfectly confident. If everything went awry, I could simply walk west until I got to Germany or something. I've done less likely things. I hitchhiked to Brazil. I drove and pushed an old Peugeot across the Sahara. I can go anywhere as long as people in uniform don't get in my way, which, I've learned, they tend to do. There's always a way around most things, but not with a hundred pounds of luggage.

Stage One of this trip — the arrival of the taxi — comes off without a hitch, and it delivers me to the station in plenty of time. Figuring out where I have to go from there involves a hike of several miles, to ticket windows, arrival/departure boards, platforms, gates, and eventually, since I can't ask anyone anything, to the Intourist office where all foreigners are supposed to go anyway. I figure somebody there knows English, but while I'm waiting around it occurs to me that maybe Intourist isn't supposed to know I'm traveling with a regular citizen's ticket, which I got from someone whose name I'd just as soon not mention. Tourists are supposed to pay a lot more, and probably in hard currency. I decide it's not worth the risk of exposing my intentions.

Whispering the phonetic sound of the words on my ticket, I figure out which platform I'm supposed to be on. A sign there seems to indicate that the train to St Petersburg might leave from that vicinity. To an intelligent-looking gentleman I utter some grunts that I hope are an inquiry about whether I'm in the right place. His thumbs-up sign seems to indicate I am.

Stage Two goes well as I make my way to the right compartment of the right car of the right train. And I'm in luck. Sharing my compartment are half a dozen girls, all about thirteen years old. Through the miracle of vodka and sign language, I learn from a man accompanying them that they are members of a judo team and he's their coach. One of the girls supposedly speaks a little English, but all I can get out of her is "Halo." She's terribly embarrassed, and her friends are all but giggling themselves to death. The coach solidifies our friendship with an unopened half-liter bottle of Stolichnaya on which he writes his address, in case I'm ever in Lvov.

They all get off the train within a couple of hours and are soon replaced by an elderly couple. I give my lower berth to the woman, explaining by pointing to the word for "Easy" in my little phrase book.

Twenty-four hours later, we pull into St Petersburg. I wait for the whole train to empty before manuevering my suitcases down the car and out to the platform.

What a miracle to hear someone speak my name.

It is Luba, a young woman whose name, it turns out, means Love. Standing on the platform, exhaling puffs of steam, the collar of her coat turned up, her hair a beautiful light blonde, she looks like a cliché who just stepped out of a spy movie. Or maybe she's an angel. She says, "I am here to help you."

I doubt I've ever been so happy to see someone I didn't know. Elena has sent her to rescue me. They met at a Moonie conference, though Luba, too, holds the conviction that it isn't likely God would send an overfed Korean to save mankind.

Luba guides me to the taxi stand outside. The driver first in line wants three hundred rubles to take us to Ana

Isabel's dorm. Luba says the price is absurd, at least ten times the standard rate. She asks another, but he says he can't go against what the first guy said or he'll be in trouble. "It's a mafia," she says to me, and goes off to find someone else. But mafia representatives follow her around, telling others what the price has got to be because it's for a foreigner.

Three hundred rubles is about two bucks. I can afford it, but there's a principle involved. I don't know how hard it will be to stick to this principle, nor does Luba know how willing I might be to spend that much money. She wasn't sent here to oversee my exploitation. We stand on the sidewalk in front of the station for a while, trying to think of something. The mafia holds a quick conference down at the corner, and one of them comes over to say a hundred and fifty would be enough. Luba looks up at me, balancing bubbles of tears in her light brown eyes. "Do you think it's okay to go with him?" she asks.

"I don't know," I say. "Is it okay with you?"

"I don't know."

"It's okay. Let's go."

So we tuck my bags into his trunk. Luba takes the back seat, I take the front. The driver gives us a nice tour of some of the most beautiful streets in the world. Half their charm lies in the lack of commercial signs. No neon, no advertisements, no color besides the muted tones of the buildings that Peter the Great had built. The city, I'm told, has more bridges than Venice, and each is of beautifully wrought iron.

Suddenly Luba is giving directions and pointing all over the place: *turn here, no, there . . . yes, wait, no, go right, go around that circle, down that street, here, yes, pull over . . .* In the middle of it she quickly says to me, "My friend lives here."

We stop at the curb of a side street. I get out, but Luba waits until the driver shuts off the engine and comes around to the trunk with the key. I pay him off, toss in a little extra, shake his hand and wave good-bye.

"*The mafia,*" Luba says, her face wet with tears. "They were following us!"

Apparently they had piled into their cars as soon as we left. Luba noticed that the same cabs were behind and in front of us for a long, long way. While I was gawking at the beauty of the city like a rube, she was figuring out how to shunt us off onto a safe street. She didn't tell me because she didn't want me to worry.

Well, don't I feel like a fool! Johnny Adventure almost loses his life to murderous cab drivers but is saved by a double agent from the Unification Church. What would I have done in her absence? Probably died on the outskirts of town, beaten to death with tire irons.

That settles it. From now on I'm not leaving home without a gun. I don't need any tremendous caliber, just a little something, loaded with birdshot perhaps, to give an attacker a quick handicap. Either that or keep Luba with me forever to foresee danger and think of what to do.

We catch another ride to Ana Isabel's dorm. Vali, her husband, is there, but Ana Isabel is in Portugal for the next week. My old visa is still in somebody's drawer at the university. By later today Vali hopes to know more. Meanwhile, there's nothing I can do about it, and he's sure everything will turn out all right.

The dorm doesn't seem to have shower facilities; I have a feeling the sink in the bathroom is where they wash. I'm too embarrassed to ask, and I don't want to ask Luba to wait around while I unpack and take a bath in a sink and change clothes, so she and I head straight for the Hermitage. It's been over a week since I've bathed, and even longer since I washed my hair, which by this time is slick and shiny with oil. For all I know, it's a little radioactive, too. But seeing the Hermitage is more important than taking a bath.

But the Hermitage is closed for Christmas. So we traipse around, check out a little crafts market where we get accosted by sleazy individual after sleazy individual trying to sell us military watches, matryushka dolls, genu-ine icons of the twelfth century, plastic amber chess sets. I buy two dolls and a chess set and would buy more if the little beezneezmans didn't bug me so much.

Luba takes me to a little restaurant for chicken and juice and something called "seafood," a kind of mayonnaise with a vague taste of low tide. Then we go to her room for tea. She shares a communal apartment with half a dozen other people, each with their own room and a kitchen for common use. This is in a building where Fyodor Dostoyevsky used to live. Luba loves Dostoyevsky, has loved his work since she was about twelve and first realized how deeply a writer can probe the human psyche. I myself don't remember *ever* reading Dostoyevsky and sure as hell not when I was twelve. When I was twelve I was reading about PT boats and volcanoes.

Luba's furnishings consist of a bed, a small dresser, a very old cassette deck, a lamp, a few posters, a couple of small dolls and some tea. She brews me up a cup and then another cup and then more.

Her boyfriend shows up. He's young, clean-cut, quick-witted, and, like everyone in an honest business, slender. He's started his own little newspaper. Like Luba, he's been goofing with the Moonies, and he does a good job of imitating one of them calling his flock together for a quick prayer. We have a good time messing with English, and Luba giggles like crazy.

No good news back at Vali's except that he has set a crack team of Lebanese friends to the task. They are unexcelled at pulling rope, bending rules, grabbing ears, buttonholing the powerful. Tomorrow they will take my passport and some photos to a certain police station and convince them to issue me a visa. I give them twenty dollars to use as a bribe if necessary.

The next morning I take the subway as previously instructed and meet Luba where previously arranged. She is late, so I pace and fret until a raggedy kid comes up and hands me a frayed slip of paper. Crude handwriting in English, faded to bare legibility, says that he is an orphan, would I please give him some money so he can eat.

Well, on a principle I established many years ago, I

don't give to beggars. Period. It doesn't solve their problem and only encourages further nonproduction on their part. Besides, I've heard too many stories of scams, too many beggars with Cadillacs and homes in Florida. I'm not sure what makes me think this kid's got a Cadillac in Florida, but who knows?

The kid looks at me as though I'm too stupid to comprehend that he's a starving orphan. He won't take *nyet* for an answer. He hangs around and hangs around until Luba shows up, reads the note and hands him some kopecks — fractions of cents. She says she can tell he really is an orphan. There are a lot of them, and with the economic crisis the country is going through, they aren't fed much, and during their teens they're expected to disappear. So then I feel extremely stupid, not to mention guilty. My beggar principle goes under immediate review.

The irony rubs in with full force when we get to the Hermitage. There's a long line of schoolchildren waiting to enter. A guard, a uniformed kid with a rifle, says it will take an hour or two to get in. Even though I wouldn't give an orphan a penny, I slip this soldier two dollars so we can get in ahead of everyone else. Inside, they have a special price for foreigners: about a nickel. If you want to take a camera with you, it's another nickel. A video camera costs yet another nickel.

I don't know what to say about the Hermitage. I can certainly say it's the most impressive thing I've ever seen, a record held previously by the Grand Canyon. Everywhere I look there's an object of shocking beauty. It isn't just the art. Every doorknob is exquisitely designed and crafted and worth a few moments of appreciation. It seems disrespectful to walk by anything without giving it a good look-over. Every square foot of floor and ceiling deserves space in a frame on a wall.

Luba and I part in the subway. She gives me two nice books, hardcover collections of stories by Chekhov and Dostoyevsky. I give her my Walkman radio. We hug, our arms barely going around each other's coats. I love Luba. I will never see her again.

Back at Vali's I hear the horror story of how my visa was not procured. His two friends spent almost all day under unofficial arrest at the police station where the cops were very angry that these Lebanese kids had an American's passport and the American was wandering around their country without it or a visa. As Vali tells the story, his friends used their powerful knowledge of police psychology to balance backtalk with deference to the power they faced, refusing to hand over the passport or to reveal their names or the address where I'm staying. Legally, I'm supposed to be in an Intourist hotel and most certainly not in a state-owned dormitory. Being foreigners, these kids got away with a lot, and eventually the police just turned them loose.

So the only thing to do, we decide, is try to bribe my way through customs. One kid, who speaks excellent English, French, Russian and Arabic, says it'll be easy. He knows someone who did it. The price of egress: fifty dollars. The trick is to get by the guy who searches your suitcase, who by all rights has nothing to do with your visa. It's the KGB guy in the booth who has to approve your exit. He's easy because he's stationed in a private spot where you can talk man to man.

So off to the airport I go, arriving at eleven p.m. for a flight at six-thirty a.m. I honestly expected the international airport of the Soviet Union's second largest city, Russia's Gateway to the West, to be open all night. It isn't. I have to sit and wait on a concrete bench, sipping brandy and hoping the mafia doesn't come along. An hour or so later, four guys come and sit around drinking vodka and smoking and peeking in the windows to see if anybody's in there. Then everybody else shows up, a hundred-odd people standing in the cold night until three a.m.

As the rough line draws into a dense crowd at the door, someone comes to practice his English, asking me where I'm from, why I'm here, etc. I show him my National Writers Union press card, which I have in my pocket as some kind of defense against whatever might happen inside. The man snorts and says, "It's a dangerous profession. Don't tell them."

Here's how the St Petersburg airport is set up. As soon as you step in you're up against the ticket counter. You can't go any further without a ticket, a passport, a visa. I talk the ticket lady out of the visa part. The next step is just beyond the counter — the customs guy. He X-rays everybody's suitcases, then opens them for a detailed paw-through. He must have known I was coming because before my stuff gets through the X-ray machine he's demanding my visa. He's real hyper, as if he's in a hurry or has to go to the bathroom or something. I tell him, sorry, no visa. That gets him all excited. He plows through my passport looking for something of interest, but since it wasn't stamped on my way out of the US or into the USSR (which was still in existence when I entered in December), there's no evidence of anything in there. Hyped up though not quite yet really angry, he demands my customs declaration, which lists everything I brought into the country, including, theoretically, every nickel and dime, though of course I lied when I filled it out so I could sell my dollars on the black market.

My story, which I hope he'll ask for and then believe, is that my visa and declaration and all but about *fifty dollars* were stolen on the train from Kiev. I didn't report it to the police, or rather I *tried* to report it, but the theft was in Ukraine, see, not Russia, so when I got off the train, the police here didn't want to know about it. Nor would they give me a new visa.

Worst case scenario, I'll inform this card-carrying member of the KGB that it is metaphysically impossible to deny or negate the truth. Their attempts to bury it under mountains of fear and dead bodies must, inevitably, fail. And obviously they are failing. Their social structure is tattered, worn thin, showing through. Under such political and economic conditions, a visa has no validity. So why should I need one? I'm leaving. Get out of my way.

But the customs guy doesn't want to hear all that. He doesn't even ask. He wants my visa. He also wants a look in my suitcase. Ah-ha, what's this video cassette? Despite a full night of practicing lies and stories, I blurt out the worst possible answer, the raw truth: "Chernobyl."

Now he knows he's dealing with an idiot. The bastard *snickers* at me and tosses the tape to a counter next to him. He says in Russian, "What's in this cinema film can, stupid?" I don't bother telling him. He tosses that over to the counter too and digs into my suitcase with extra vigor, the damn sleuth. "Where's your visa?" he says, now mad.

"No visa," I say. "It was stolen on the . . ."

He cuts me off to fetch a higher-up, a short customs dude in a well-pressed green uniform with badges that say CCCP and KGB, two organizations which no longer exist. This guy doesn't want to hear my story either. He hails someone who might be a sort of porter and tells him to haul me the hell out of there. In the confusion, I stuff my video and film and other stuff back into my suitcase. I make a last appeal at the Aeroflot window, explaining that I'm going to miss my flight if I don't get through and the next available seat is six weeks hence. The nice lady tries to talk to a large woman in a green uniform, but her nose goes up in the air so fast you could swear she smelled something. I wave my dollars at her, saying "Isn't there *some* way I can *buy* a visa here, with my *dollars*? Isn't there some way we can settle this problem right here? We have to or I'll miss my plane. It's the last flight for the next six weeks." But she doesn't want dollars. She just says, in roughshod English, "That is your problem, not ours," and stalks off in search of bigger fish to fry.

And so there I am out in the cold, dark parking lot, three-thirty a.m. in the middle of the winter at the end of the world, thinking how wonderful life can be. Someone offers me a hundred-ruble ride back to Vali's. By a miracle, the front door of his building is open. I wake him up. It's hard for him to get back to sleep after that. Now he has a headache — me, with no place to stay, and it's all his fault, or Ana Isabel's anyway. How are the three of us going to share a one-room dorm for six weeks?

I get up early and take a cab to the central post office to send a fax home. As far as my wife knows, I should be arriving in New York in a few hours. I pay a nice lady two hundred and fifty rubles and leave a scribbled fax explaining that I might not be back until late February or early March, please advise the two universities where I teach, please find some other way to pay the mortgage, the car insurance, the heating oil company. I'm hoping the fax might arrive before the plane does, maybe even before my wife leaves on the three-hour trip to Kennedy airport. The lady says *nyet problem,* it will be there in a couple of days.

The only Aeroflot office in town happens to be two blocks from the post office, so I go to check flights. Well, there *are* plenty of Business Class seats on all the flights, so I optimistically sign up for one on Thursday of the following week, by which time a visa will flutter down from heaven. Or maybe I'll find one on the floor of the Hermitage, which happens to be right down the street. How convenient! And no line waiting this time, so I save two dollars and only have to pay the four-cent entrance fee.

Not bad, considering how much some of the chandeliers are worth.

Here, by the grace of God and the KGB, I figure I'll spend the next week. I can't think of anything else I should be doing or even could do if I wanted to. Certainly there's no leaving town. I can hang around the museum reading a little Dostoyevsky, writing in my journal, trying to see every item in the place — the latter being an apparent impossibility. I can't even figure out how to get from the Rembrandts to the Van Goghs. There are no little maps you can carry around, and the framed maps they have on posts in each major room are absolutely uninterpretable. They're complex grids of little white lines with no indication of where you are while you're looking at them and which way is, say, north. The list of the contents of each room is in Russian. You could ask the little Russian tank they have sitting on a chair near the door in each room, but she speaks only Russian and she's there mostly just to tell kids to keep their grubby mitts off this and that or to stop giggling at the nudes.

So the only way to figure this place out is to wander around as fast as you can until you gain a gut feeling of which way to turn to get somewhere. But you'd have to have the aesthetic sensitivity of a farm animal to walk very quickly very far without stopping. The only way to see this place in a day would be on a trail bike going fast enough not to let anything distract you. If it isn't a DaVinci, it's a fabulous urn the size of a phone booth. You can reach out and put your fingers on the ink of a Renoir. The ceilings are celestial, the floors all that floors can be. The windows are wonders, tall and framed in grainy wood and looking out over the Niva River or the vast cobblestone plaza of the Military School, or inward at a courtyard. Through the shimmer of the old glass, the courtyard looks like a painting by Monet, though in dark, cool stark tones that the impressionists didn't see in France. The trees are purple brown and so are the stone pathways among them, all burnished by melting sleet. The ground has brush strokes of light, wet snow over it. A man with a cane limps into the scene, fiddles with something down low near a tree, then stands as if in thought.

I case the joint, looking for nooks where I might spend the night. I see a czarina's bed that would be absolutely perfect. The best spot I can find is a roomy Egyptian coffin, the empty one next to the one with the mummy, where I could lie out of sight of the motion detector above the door. I stand there for a long time wondering how I could climb in without being noticed and whether I'd really do it if I had to.

I find a nice spot to write, a velvet chair below a fifteen-foot window in the Greek statue department. The naked, sleepy-eyed goddesses behind me are as old as Western civilization. It's a perfect place for a guy like me. I smell like a pig, having not bathed in a week and a half. But that's okay in this case. It keeps people from asking me if I speak English or want to change dollars or buy counterfeit tickets to the ballet or a small Picasso in a solid gold frame. One haggard man offers to sell me his military papers, ID card, and other official documents. The statues don't mind if I sit here. I guess over the course of the last couple thousand years they've seen it all; to them, the Hermitage and I are nothing.

A little before closing time, I return to Vali's via subway, getting good and lost along the way. By the time I untangle myself, painstakingly deciphering signs and sometimes going to the next station just so I can figure out the station I was at before, it's well after dark.

Vali arrives at the same moment. He and his friends have managed to get hold of my old visa, but they say my only hope is the US consulate, which doesn't sound hopeful to me. They'd have to basically lie to the Russian government, telling them I was working there or something. But I plan to go there tomorrow and throw myself on their mercy, explaining that if they don't get me out of there and soon, they'll eventually have an emaciated radioactive cadaver to deal with — surely every consulate's nightmare.

Preparing for the worst, Vali, quintessential Lebanese, finagles a dorm room for me on the seventh floor. His buddy is to pick up the key from the *komandante* who runs the building. I'll pay her an amount to be determined later. It's a good room, with a bathroom. I can't imagine how I'll find something to eat for the next week, but a dorm with a bathroom is better than being out in the cold or sleeping in an Egyptian sarcophagus or sharing a one-room apartment with newlyweds.

Vali and buddy give me a ride to the US consulate on their way to the airport. (They're going to Moscow to pick up Ana Isabel, who's supposed to be laden down with a hundred pounds of food.) A staffer at the consulate, Janine, saves me. She asks only that I request an extension of my old visa at a Russian office that just happens to be right around the corner. I go there, wait a while, get my *nyet*, go back to the consulate, leave a note for Janine, and go have a little lunch in a little restaurant that actually has food on the menu, albeit not much. The soup looks the safest, and indeed it's pretty good, a borscht-like substance in the kind of quaint little crock you'd find in a quaint little upper west-side restaurant in New York, at a price no doubt barely exceeding a quarter of a cent.

Meanwhile Janine types up the letter I need to deliver to the Russian office. I take it on over and, with a noticeable smirk, hand it to the lady who a few moments earlier had said *nyet*. In the name of diplomacy she does what she must do. I can't help but think that I have 30,000 nuclear weapons and as many tons of Marines standing behind me. I think of all the tax dollars I've dumped into the military budget and decide they were worth it.

Then I manage to find the right subway to the right subway to the wrong station to the right station to the Aeroflot office, where they cheerfully give me a ticket for *thatvery next morning* at four-thirty a.m.

Ain't it amazing? You pull the right string and everything flushes.

Back at the dorm, of course no one's home. I repack for customs. I leave Andrew's film on Vali's kitchen table with instructions that they can keep it, sell it, show it, or send it back to Luba to give to Elena to give to Ljudmula to give to Andrew. I rip the covers off any magazine or book that has the word Chernobyl on it. I put a diskette of my notes in my money belt. I stuff hard copy of my notes down the back of my pants. I rip apart my camera bag and create a false bottom where I put suspicious papers and photographs. I write "UN Conference" and "Sobchak Interview" (Sobchak's the mayor of St Petersburg) on all film canisters and tuck them in various jacket pockets and the space in my jacket collar where the hood rolls up. I keep the video tape in my coat pocket. I fold a fifty-dollar bribe into my shirt pocket. I put on a conservative necktie and my professor's jacket.

I take a shower in a bathroom upstairs.

I'm back at the airport by midnight. My KGB buddies are there, too. At the ticket counter, somehow my ticket gets bumped down from Business to Tourist class. I don't like this. Economy is full, Business empty, the flight long and hard. But I don't argue. I send my suitcases through the X-ray machine. The KGB kid monitoring my contents asks me about the typewriter he can see in there. He asks me if I bought it in Russia. I give him a look. I need say no more. He passes me along. I zip through the rest of the process as if lubricated. They don't ask for my visa, don't even search me. Ain't it amazing? It must have been the shower.

(Either that or the CIA. My wife, not having received my fax, began to worry and called a partner of mine who, unbeknownst to her, had been in Vietnam with someone who now is with a major government organization in

Langley, Virginia. It so happened that that person's job involves monitoring the airports in Russia to know who is going in and out. Apparently he found out or figured out that the first time I tried to go through, they would be waiting for me. Apparently he told them I was small potatoes, so the next time I tried, it was easy. Apparently I had a code-name: Shitferbrains.)

I push my luck — a sneak attack on Business Class. I use my Swiss knife to carefully scrape the "one" from in front of the "seven" in my seat number. It looks obvious, though, so I put the pass on the floor and grind my boot against it, pretty near mutilating it. Now it looks as though I could conceivably have misread it. I just hope they'll figure anyone in a plaid necktie and a jacket with leather elbow patches has *got* to be Business Class. But I'm sure it won't work, especially if I have to lie to back it up. I'll do better feigning dumb. They don't call me Shitferbrains for nothing.

The first test is on the runway on the way out to the plane, which is the only vehicle out there in the snow-blown dark, a hulking, overfed pterodactyl a hundred yards from the terminal. It's quite a scene, an IL-86 on a runway at four o'clock in the morning, overweight, serious, cold, grim with determination and perhaps a bit of fear. Army guys stand near the back of the stairs and under the wings, their long khaki trench coats flapping in the wind. People in Business Class flash their red boarding passes and get to go up the stairs in front. Mine isn't red, so I go up the back stairs like a servant and then, like a Lebanese, wander forward toward quarters more suitable for a guy with leather on his elbows.

There's a dog down in the luggage compartment. No, wait, *two* dogs. One barks with powerful lungs. The other yipes in terror. I don't blame them. The baggage compartment is just down a flight of stairs. You have to walk through it to get on and off the plane. It's a scary place to have to stop.

There's plenty of room in Business class. We've got three or four seats each. A dog could come up here, no problem. Wouldn't bother me in the least. I could use the company. A good dog would act like me, keep his head down, quiet, unobtrusive, blending into the scenery with all his might.

Alas! Caught! A flight attendant who seems to have done her make-up with Crayolas says, "Your seat is 17f, sir. Please to go back."

So I do. But there's a kid sitting in 17f and no empty seats. I duly report the fact. Since the engines are warming up, the stewardess lets me assume 7f anew. Out the window I can see the predawn wind flapping at the coats of the soldiers who stand at attention under the wing. For joy, for joy! Good-bye, Russia. May the wind be at your back. Always.

Epilog

In early 1993, a fire broke out in an electrical shed a hundred feet from Chernobyl Reactor One. According to a Chernobyl spokesperson, the resulting fire, which took an hour to extinguish, was no more serious than a burned-out light bulb. During the emergency, no one was sure what the shorted wires controlled in the reactor, but authorities saw no reason to shut down Reactor One.

A few weeks later, some lumber inside the sarcophagus around Reactor Four inexplicably burst into flames. A few weeks after that, an oil line in the generator room of Reactor Three burst and sprayed the room with oil. Later in the year, thieves made off with two fuel rods packed with uranium.

In late 1993, with the country low on fuel, and electricity to Kiev being cut off daily, the Parliament of Ukraine voted to continue operating Reactors One and Three. They postponed a vote on whether to repair and restart Reactor

Two, and a decision on whether to continue building new reactors in Ukraine, including, perhaps, one or more at Chernobyl.

In 1995, former Minister of the Environment Yuri Shcherbak, now ambassador to the United States, said that the Chernobyl disaster had left 8,000 dead and another 30,000 with diseases related to radiation. Thyroid problems in children had risen by a factor of 80. Cancer rates were soaring and not expected to peak until some time in the next century. Thirty thousand people still needed to be evacuated and resettled. The birth rate was two-thirds that of the years before the explosion, and the death rate, one third higher than before, now exceeded live births by forty percent. For every live birth in Ukraine there were eight abortions. Among high school children, only eighteen percent of boys and eleven percent of girls are considered healthy. In a phenomenon called the Second Chernobyl Wave, people's internal levels of radiation, because of the contaminated food they are eating, have risen to the levels they had reached in 1987. A team of Japanese researchers from the University of Hiroshima, after studying 30,000 newborn babies and stillborn fetuses in Belarus in 1994, determined that the rate of birth defects in contaminated areas had doubled since 1986.

With the results of the accident growing ever worse, the U.S. Department of Energy released a study in mid-1995 finding that nine Soviet-type reactors operating in Eastern Europe were deteriorating, "raising the spectre of another accident akin to Chernobyl." The four most dangerous reactors were termed "accidents waiting to happen."

Ranked most dangerous of all was Chernobyl itself, where conditions were "in many ways worse" than they had been when Reactor Four blew up.

The latest plan is to shut down the entire Chernobyl plant and build a gas-fired plant nearby before the end of the century. The cost is estimated at $4.4 billion. No one has any idea who will pay the bill. The only thing certain is that it won't be Ukraine.